Running *WILD*
The Next Industrial Revolution

Running *WILD*

The Next Industrial Revolution

Adam Osborne

OSBORNE/McGraw-Hill, Inc.
Berkeley, California

Published by
OSBORNE/McGraw-Hill, Inc.
630 Bancroft Way
Berkeley, California U.S.A. 94710

For information on translations and book distributors outside of the United States of America, please write the publisher.

RUNNING WILD — THE NEXT INDUSTRIAL REVOLUTION

1234567890 CPCP 78765432109

ISBN 0-931988-28-4

The research editors for this book were Nora Mote and Susan Granzella.

Contents

Introduction

We are already five years into a new industrial revolution, the impact of which will rival the first industrial revolution.

Of all jobs in the industrial world today, perhaps half will be eliminated during the next twenty-five years. From assembly line to stockbrokerage, blue collar and white collar jobs alike will disappear. The jobs that do remain will make new demands on job seekers and job holders. With good planning, the net effect could be beneficial to us all: people will spend less time making ends meet and more time pursuing their lives' goals. But without adequate planning, we could be heading for a time of anguish and chaos. The misuse of computers, for example, is already an urgent problem which must be addressed immediately.

The new industrial revolution was created by

the semiconductor industry, which is centered in an area south of San Francisco known — affectionately to some — as "Silicon Valley." This name derives from the principal product of the semiconductor industry: complex electronic circuits built on tiny pieces of silicon. These circuits are called "microelectronic devices." (A non-technical description of microelectronics is given in Appendix A.)

Why should the semiconductor industry, and microelectronics, be so special as to cause an industrial revolution? Recently journalists have given microelectronics much sensational press coverage. But for once the journalists were confronted with a reality too amazing for even their imaginations, and the sensationalism has been justified. We have witnessed technological breakthroughs and price reductions that are stunning, even in comparison to the mind-numbing advances we have seen over the past fifty years. For example, a computer that filled a room and cost half a million dollars in 1959 will fit on your fingernail and cost five dollars in 1979. The electronics inside your home microwave oven would, twenty years ago, have been larger and more expensive than the oven itself.

Technological advances and price reductions have brought about a revolution because no one has been able to predict what would happen, or plan for it. The leaders of the semiconductor industries themselves constantly underestimate what their companies will be able to do. The presidents of high technology companies, who should have seen what microelectronics would do to their companies' products, were usually the last to get the message. Entire industries, such as those manufacturing mechanical calculators and watches, have been severely crippled or wiped out by competition from a quarter they never anticipated: the semiconductor industry. And with each industry that gets wiped out, a labor force loses its jobs. That is the consequence of revolution.

What can we do for the future?

Perhaps the most paralyzing aspect of the microelectronics industrial revolution is the inability of law makers and sociologists to cope with what is occurring, because no one knows where to begin. We can legislate, but we cannot anticipate what the real impact of any legislation will be.

Nor can we turn for counsel to the senior management of high technology companies. Their track record as predictors of the technological future has been nothing to brag about. And my sympathies are with these senior managers. When writing the first draft of this book, I predicted a number of events to occur within the next ten years; further research revealed that most of these events had already occurred. Nevertheless, it is possible to make some technical predictions regarding events that microelectronics will make possible. But it is almost impossible to predict the interaction of technology, economics, and human nature. Many changes which are technologically possible will be blocked by a skeptical or hostile population. For example, automats have, for fifty years, given us the ability to eliminate waiters from restaurants. The Horn & Hardart chain of automats uses coin operated machines to dispense food. In the future, you could punch numbers into a control panel and have a machine prepare exotic meals to your command. But I suspect people would reject this alternative to waiters, whose profession will survive and even thrive in a new technological world. On the other hand, even if a society frowns on robots taking over production line jobs, economics will make this development inevitable.

It is relatively easy for me to predict what microelectronics can do, but it is hard to estimate the social consequences of these predictions. Perhaps that is because I am an expert in microelectronics, not in the social sciences. Therefore, in this book I limit my ambitions to describing technological events that have occurred and forecasting events that I believe are possible. I attempt to weigh economic and human considerations against technological feasibility, but only as a casual observer. My sociological and economic conclusions must be evaluated as the conclusions of a layman.

I discuss the impact of the microelectronics revolution on a number of common trades and professions; of course, I cannot cover every trade and profession, but parallels are easily drawn.

1
Roots

The problem with politics is that few of us are wise enough to foresee the many ramifications of any legislation.

In 1960, President John F. Kennedy declared that within the decade an American would walk on the moon. One did, but history may well be more impressed by unanticipated consequences of President Kennedy's declaration. Unwittingly, he triggered an industrial revolution.

The space race artificially created hundreds of unstable, high technology companies. In addition to producing wondrous products, these companies produced employees trained to become entrepreneurial gypsies. The entrepreneurial gypsies took the wondrous products to the marketplace, and in the process they became exceedingly rich because, strange to say, established companies were ineffective competitors. The exploits of the early entrepreneurs are now highly visible; imitations abound, and the consequence is a new industrial revolution.

Let us examine this scenario as it unfolded.

Most observers agree that the space race of the sixties propelled the microelectronics and semiconductor industries into an era of rapid development. These industries manufacture tiny pieces of silicon, smaller than a child's fingernail, each providing more electronic logic than once filled a room. But for the public at large, the only tangible and immediate benefits of the space race were the spectacular television shows covering rockets leaving the earth and men walking on the moon. (The space race also gave us sundry necessities of the good life, such as Teflon* coated frying pans, and plastic bags you can put in the oven.)

To put a man on the moon, the U.S. Government showered high technology companies with contracts to research and build everything conceivable. Many new companies were founded solely to service U.S. Government contracts because there was simply too much work and too few companies to handle it.

In 1978 Standard and Poor's Register listed 147 companies in the "Semiconductor and Related Devices" category. In 1970 only 85 companies were listed in this category. In 1960 Standard and Poor's Register had no such category.

The American Electronics Association listed 728 member companies in their 1978 directory. 282 of these member companies did not exist in 1970. Only 152 companies were founded prior to 1960.

* Teflon - Dupont's Registered Trademark

The U.S. Government tried very hard to control the profit in its contracts. Strict rules governed the amount of money that could be spent on management overhead, employee benefits, and overall profit. These rules kept politicians happy, but as usual, human ingenuity won out and government regulations were circumvented.

Most of the money in any government contract wound up paying salaries (for scientists, engineers, and other supporting employees). Companies receiving contracts (contractors) were told to calculate their profits as a percentage of total salaries paid. Therefore, as salaries increased, or the number of employees on a job went up, so did profit. A government contract might allow the contractor to charge twenty percent of "on-the-job" salaries to cover management overhead and profits. Suppose a contractor in the mid-sixties had two junior engineers on a job, earning $20,000 a year between them. The contractor could then charge $4000 over and above salaries, to cover administration expenses and profits, and the government could easily check that this amount was not exceeded. But the government could not so easily check when the contractor decided the job was too complex for two junior engineers. Suppose the contractor claimed a need for five engineers, two secretaries, three draftsmen and a technician, between them earning $150,000 a year. Now the contractor could charge the government $30,000 to cover overheads and profits — more than the total job would have cost using two junior engineers.

That is precisely what many contractors did. They padded the payroll and they inflated salaries. While the contracts lasted, this was great for everyone. Employees certainly liked the idea of big salaries, and the overstaffing gave engineers and scientists time to learn new subjects, in a way that would have been impossible had projects been understaffed and over committed.

But this rosy picture ended abruptly with the end of the government contract. There was nothing in the government contract to cover job retraining. Worse, contractors were left with bloated teams of employees, earning excessively high salaries. So employees were laid off the day after the contract ended. Scientists and engineers frequently changed jobs four or five times within the first ten years of their careers, hopping from one contract to the next. Soon they were attuned to seeking instant rewards rather than working for long term career goals. In contrast, a chemical engineer who graduated at the same time might have taken a job with an oil or chemical company; his first job would likely last thirty years or more, until he retired.

The job-hopping graduates of the space race faced potential unemployment, but they got lucky. In their space age jobs they constantly saw new technological innovations with big commercial possibilities.

Imagine a group of electronic engineers sitting around a coffee table, discussing the rocket control computer they have just built. One engineer points out how easily this computer could be redesigned and built, for very low cost, to be sold industrially. The electronic engineers have two options: they can take their idea to company management, or they can quit their jobs, start a new company, and build the computer themselves.

A group of chemical engineers talking over a similar discovery (perhaps a new synthetic fiber) would, without doubt, take their discovery to company management. The thought of leaving secure jobs and vested pension rights to start a new company would be incomprehensible, and their employer would surely sue them for stealing secrets.

But space race engineers did not see it that way. Probably none of them had been in any one job more than two years, and for all they knew, they might be fired tomorrow, the lot of them. So why not take a chance? They had little to lose and much to gain. What is more, their employer was probably too broke to sue them for stealing secrets.

That is how thousands of new, high technology companies began.

A surprising number of these new, high technology companies have been very successful, making their founders exceedingly rich, because newness has been a big advantage for a company when taking high technology products to market, and opportunities have abounded.

Why is newness an advantage? New companies fare badly when they face established competitors, selling well-known products into existing markets. On the other hand, established companies fare badly against new competitors in emerging high technology marketplaces because established companies can rarely move fast enough to take advantage of opportunities as they occur.

And when it comes to opportunities, the era of the sixties and seventies has been the equivalent of the California gold rush in electronics. New opportunities fell, and continue to fall, out of the electronics industry like gold nuggets washed by river water out of the mountains. The first passersby pick up these gold nuggets of opportunity, since no one has been there before to do it.

And thus, the new companies, which were created to handle government contracts, themselves spawned a second generation of new companies, which took the inventions and technical expertise of the space race into commercial and industrial markets.

These new, second generation companies had little impact on the population at large; generally they built small computers and electronic equipment that did much to change the way business did business but sold nothing directly to the public.

Consider the "minicomputer" industry. Beginning in 1962, a number of companies started manufacturing small industrial computers, generally called minicomputers. The name "minicomputer" derived from the relatively small size and reduced capabilities of these new products. Minicomputers were supposed to go into machines rather than support data processing applications. Today the three leaders of the minicomputer industry are Digital Equipment Corporation, Data General, and a division of Hewlett-Packard. Their sales between 1962 and 1979 increased dramatically.

You have probably heard of Hewlett-Packard; it manufactures a wide range of products. But Digital Equipment Corporation and Data General, despite their phenomenal growth, are unknown to the population at large. Why?

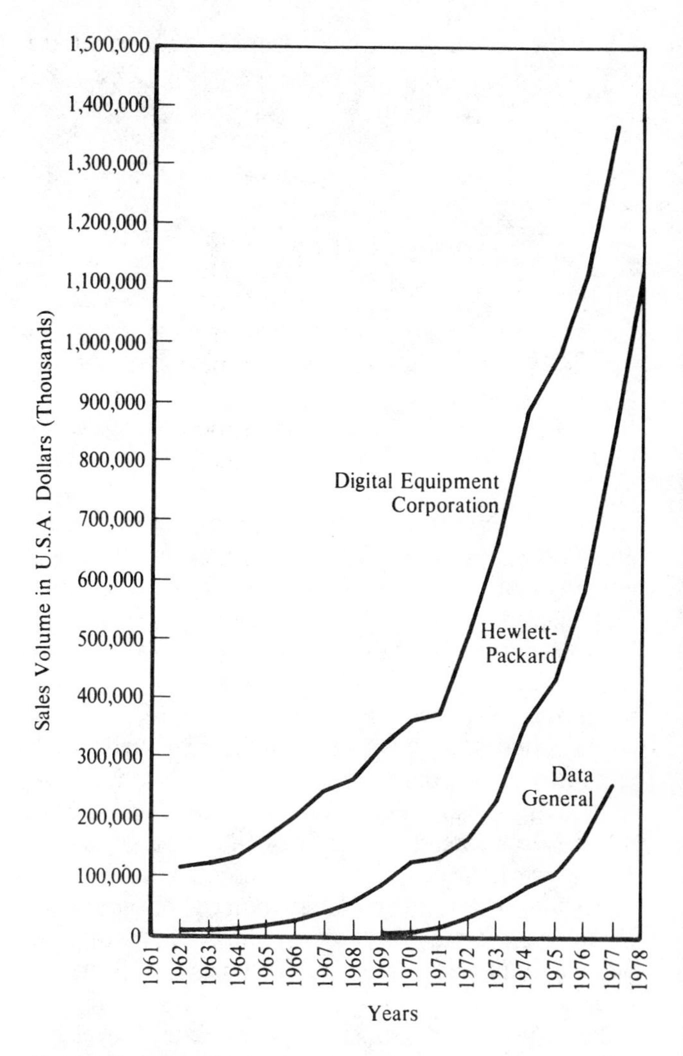

Graphs of net sales for the three leaders of the minicomputer industry, Hewlett-Packard, Digital Equipment Corporation, and relative latecomer Data General. (Sources: Hewlett-Packard Annual Reports (1962-1967), Data General Annual Reports (1969-1977), Digital Equipment Corporation Annual Reports (1962-1978).)

Photo courtesy of Texas Instruments, Incorporated.

The inner workings of an inexpensive pocket calculator. The minuscule microelectronic device that performs all calculations is inside the package indicated by the arrow.

Because their computers revolutionized industrial automation, which has low visibility, while IBM was revolutionizing data processing, which has high visibility.

You are unlikely to be impressed by the profound impact that minicomputers had on the machine tool control industries. This and similar developments of the sixties spectacularly reduced manufacturing costs and changed job classifications within many companies, but all the general public saw was steady prices or prices rising more slowly than the inflation rate. Electronic calculators brought microelectronics to the general public for the first time. Soon everyone had his or her own calculator and quickly forgot how to do mental arithmetic.

In 1968 I was working for the Shell Development Company in Emeryville, California. I remember my department buying a mechanical calculator manufactured by the Friden Division of Singer. This machine cost approximately $1200, weighed forty pounds, and took one or two noisy seconds to execute any calculation. Today you can buy a more powerful electronic calculator at the supermarket for $10, and you can slip it into your pocket.

Have you ever opened up one of these $10 calculators to see what it contains? You should do so; it is worth the $10 you may lose if you cannot put it back together again. Inside you find a piece of plastic, with a small electronic "insect" embedded in it. This piece of plastic connects an inexpensive set of keys to little lamps that display numbers.

The electronic "insect" is a microelectronic device, the product of the microelectronics and semiconductor industries. The electronic "insect" contains a small piece of silicon which holds all the electronic logic for the calculator. The "insect's" legs are the electrical connections. The piece of silicon is called a microelectronic circuit.

The electronic calculator you buy is no longer manufactured by the Friden Division of Singer; they are no longer in the calculator business. Monroe, NCR (National Cash Register), and other prominent manufacturers of mechanical calculators are also non-entities in the new electronic calculator market. Today your calculator is manufactured by Commodore, Casio, Hewlett-Packard, or Texas Instruments, to mention a few. Of course, not every company that started manufacturing electronic calculators did well. As all good revolutions do, the microelectronics revolution produced casualties. For every successful company, there were probably ten that tried but never made it. Do you remember such names as Bowmar or Litronix or

Photo courtesy of Intel Corporation.

The microprocessor is housed in a Dual In-Line Package (DIP). The legs of this "insect" are electrical connections; the body of plastic contains the microelectronic circuit.

Eldorado? These are just three of many that tried to make it in the calculator business but failed.

Electronic calculators were the brain children of companies that saw a need and set out to fill it. Companies guessed that people would buy cheap calculators in high volume, so they contracted with semiconductor manufacturers to build the microelectronic circuits that could make electronic calculators possible. Engineers and scientists who graduated from the space race founded many of the semiconductor companies that built the first microelectronic circuits for calculators. Opportunists all, they quickly decided to build calculators themselves for the public; after all, a calculator is little more than a piece of plastic surrounding a microelectronic circuit. Then the scramble was on. Everyone was competing with everyone else, and your best customer was likely to show up as your strongest competitor. Prices tumbled until Texas Instruments arrived and forced calculator prices as low as they could go.

Why stop at electronic calculators? Their appetites now whetted, semiconductor manufacturers looked around for other opportunities, and they quickly found watches.

The electronic watch first appeared in late 1969. Seiko, a Japanese watch maker, offered a few quartz crystal electronic watches, in heavy gold cases, for $1250 each. These watches were sold in Japan only and were aimed at the "prestige" market.

Watch manufacturers dominated the early electronic watch market, keeping electronic watches in the high priced market segment. In 1970-1971 there were three manufacturers of electronic watches: Seiko of Japan, the Hamilton Watch Company in the U.S.A., and the Swiss manufacturers' research group, Centre Electronique Horloge S.A. (CEH). These three artificially restricted their combined 1971 electronic watch output to 12,000 pieces, selling at prices ranging from $650 to $2000 each.

The earliest electronic watches were traditional in appearance. They had hands and numbers like their mechanical predecessors. Today we call these "analog" watches. "Digital" watches, in contrast, display numbers. There are two types of display: Liquid Crystal Displays (LCD) and Light Emitting Diodes (LED). LCD's use very little power, so you can leave the display on indefinitely. But you cannot see an LCD in the dark. LED's use too much power to leave on continuously, so you must press a knob to display the time when you want to use the watch, but LED's are visible in the dark.

Seiko, Hamilton, and CEH relied on U.S.A. semiconductor manufacturers for their electronics and their displays. Electronics first came from Texas Instruments, Intersil, RCA, and Motorola; many others joined the pack later. Displays first came from Hewlett-Packard, Electro/Data, Texas Instruments, RCA, and soon, hoards of others.

It is hard to piece together what happened to the electronic watch business beginning in 1972. What is clear is that the watch manufacturers, who had hoped to control this market, lost all control. Most of an electronic watch — its working parts and its display — came from semiconductor manufacturers who do not know how to think in terms of low volumes and high prices; they understand only high volumes and low prices. Some industry spokesmen claim that in 1972-1973 semiconductor manufacturers pleaded with watch makers to enter the high volume, low cost electronic watch market, and when the watch makers refused, the semiconductor manufacturers stepped in out of necessity. Others claim that semiconductor manufacturers never gave the watch makers any choice. But in a replay of the electronic calculator story, semiconductor manufacturers, with ill-conceived haste, decided that the watch industry was theirs. The first semiconductor manufacturers to market their own watches included Microma, Litronix, Fairchild, and National Semiconductor, but their products were mostly shoddy, with sticky switches, ill-designed cases and wrist bands. Nevertheless they survived for a time because of their low prices.

Texas Instruments was one of the last entries, but it took the time to prepare a well-designed watch. It has the lowest manufacturing costs of any semiconductor manufacturer, so today it dominates the low cost watch market.

Hewlett-Packard was also a late entry into the electronic watch market, but its products, in Hewlett-Packard tradition, are high quality and expensive.

Texas Instruments survived in the electronic watch business by going after the low-end, inexpensive market segment. Its brutal price cuts were referred to by Richard Stadin, watch marketing manager for National Semiconductor, as "the most expensive advertising ploy we have ever seen."* National Semiconductor got out of the watch business shortly thereafter. So did Litronix and Microma.

Fairchild lost more than twenty million dollars on electronic watches before packing it in in January of 1979.

Remaining in the electronic watch market today are Texas Instruments, Hewlett-Packard, and Casio, all primarily electronics industry companies, plus Timex, Seiko, and Bulova, together with numerous smaller companies that are primarily watch makers. Watch makers are, after all, better watch sellers, and the semiconductor industry has better things to do.

But what is the future of the electronic watch? Fairchild estimates that in 1978 more than thirty percent of all watches sold were digital electronic, with an additional ten to twenty percent quartz electronic. Many industry spokesmen guess that, within ten years, more than eighty percent of the watches sold will be electronic.

*"$19.95 Watch Coming from T.I." *Electronics,* January 22, 1976.

Why did electronic watches and calculators appear so suddenly and cause such turmoil in their respective industries? Why did they cause revolution rather than simple product evolution?

The answer is a surprising one: management in most large companies does not intrinsically understand its product. Rather, management understands its customer base and how to sell to it; it understands its manufacturing process and how to operate efficient production lines.

For an example, look again at electronic calculators. Any engineer could have built an electronic calculator back in the early fifties, twenty years before electronic calculators eliminated the mechanical calculator industry. Hundreds of enterprising engineers probably did build electronic calculators, as labors of love, in those early days. But what they produced would have been too bulky to be practical, too expensive to sell, and too unreliable to use. Nevertheless, as electronics advanced the day came, in the early sixties, when an electronic calculator could be built — expensive, yet saleable, unreliable, yet usable. Wang Laboratories was the first company to produce such a calculator; its "LOCI" (Logarithmic Computing Instrument) was introduced in January 1965. The 300 Series, which followed in October of the same year, was priced in the $2000 to $5000 range. Wang Laboratories sold its calculators to large companies, which bought them for their engineers. Wang Laboratories was in competition with mechanical calculator manufacturers such as Friden and Smith-Corona Marchant (SCM), whose salesmen understood metal gears and electric motors. They did not understand electronics and could no more enter the electronic calculator market than they could enter the automobile market.

At that time, Wang Laboratories was a brand new company, learning its first business. Unfettered by the constraints of prior practice, they quickly learned how to sell engineers electronic calculators priced between $1000 and $10,000.

But Wang Laboratories, in turn, ossified. Soon its $1000 calculator could be built for $100. But the $100 calculator sold to a much broader market than the limited engineering customer base Wang Laboratories understood. Wang Laboratories never learned how to sell cheap calculators to a mass market; it only knew how to sell electronic equipment costing $1000 to $10,000, preferably to engineers. Hewlett-Packard picked off the $100 calculator market. Today Wang Laboratories is out of the calculator business. It sells computers and word processing equipment in its traditional price ranges — and is growing at a phenomenal pace.

Wang Laboratories has been very successful, because it identified a customer base and it learned how to sell to this customer base. Not everyone was so successful. Many companies built viable products but never learned how to sell them, or to whom. This was particularly true of the watch industry. As previously stated, Litronix, Microma, and Fairchild were three of the earliest digital watch manufacturers — and three of the least successful. They were unsuccessful because even though they knew how to build digital watches, they never learned how to sell them. Fairchild and Litronix have been particularly inept in this respect and singularly slow to learn.

Watches and calculators have one thing in

common: they represent products with a previous history. But video games introduced a whole new twist: a product with no prior history.

The first video game was "Odyssey," invented by Sanders Associates, of Nashua, New Hampshire. Magnavox, the well-known television manufacturer, obtained exclusive rights to this game. It started marketing the game in January of 1972, with a lack of vigor that underwhelmed us all; few people even remember Odyssey, and fewer realize that Magnavox was even in the video game business, let alone that it was one of the pioneers.

Nolan Bushnell and his company, Atari, turned video games into a consumer product. Bushnell graduated from the University of Utah in 1968 with a B.S. in Electrical Engineering. He spent a few years with Ampex Corporation before designing "video pong" and founding Atari Corporation. The story of Nolan Bushnell, Atari, video pong, and the subsequent emergence of microelectronic-based games represents, in microcosm, the entire revolution that is occurring.

You might guess that a company such as Parker Brothers, Mattell, or Brunswick Corporation, all established in the games business, would have popularized electronic games. But the electronic games industry finally got started with video pong, invented by an engineer with neither the financial resources nor management background suited to such a task.

You might have expected that electronic games, having been invented, would be picked up by a company established in the games business. But Bushnell found it more attractive to start his own business.

You might have expected a new, small company, marketing an obscure invention, to wither in the face of established competition, or at best to establish for itself a specialized niche, marketing a product that very few people want to the very few people who do want it. But instead, Bushnell and Atari spawned an entire new industry. In 1976 Nolan Bushnell sold Atari to Warner Communications, and Bushnell himself, in his early thirties, became a multi-millionaire.

The story of video pong, Nolan Bushnell, and Atari is not uncommon; it is the rule rather than the exception in the microelectronics industry.

The space race established an environment of opportunism, a generation of chance-takers, and the possibility for success.

The inability of established companies to adapt to changing environments reduced their ability to compete in new markets. Instead, established companies bought out their new competitors — once the new competitor became a growing concern. And that is how the entrepreneurs who founded the new companies became wealthy.

This combination gave rise to hundreds of success stories, each of which might make an interesting book in its own right.

But far more interesting are the future consequences of the environment that has now been established. If something is possible and many people try it, ultimately one of them will succeed. And many people will try. Nolan Bushnell made his fortune, via Atari, within five years; so did hundreds of others. There were, of course, many who failed, but the success rate has consistently been high enough to keep everyone trying.

That is why the new industrial revolution is

upon us, and it is beyond control. We have turned loose a multitude of entrepreneurs. By the thousands they are "trying it," and anything electronic that can happen will happen — provided it is even slightly possible.

Calculators, watches, and video games each demonstrate a different aspect of microelectronics/semiconductor industry resourcefulness. Calculators illustrate the industry's ability to build circuits on demand. Watches illustrate the industry's ability to spot existing opportunities for microelectronic technology. Video games demonstrate an ability to create new opportunities by building products that previously could not have existed.

These three capabilities of the microelectronics/semiconductor industry combined to generate the "microprocessor." A microprocessor is the "central processing unit" or "brains" of a computer, manufactured as a single microelectronic device. This single microelectronic device may cost less than ten dollars, yet it is equivalent to computers that cost half a million dollars twenty years ago.

Today the microprocessor fits on your fingertip. Five years ago this tiny circuit would have been the size of a briefcase, and ten years ago it would have filled a desk. The briefcase-sized machine cost approximately $3000, the desk-sized predecessor, perhaps $150,000.

The microprocessor, more than any other microelectronic invention, is responsible for the microelectronics industrial revolution. Calculator, watch, and video game circuits are dedicated to a specific task — making a calculator, driving a watch, or playing a game. But the microprocessor

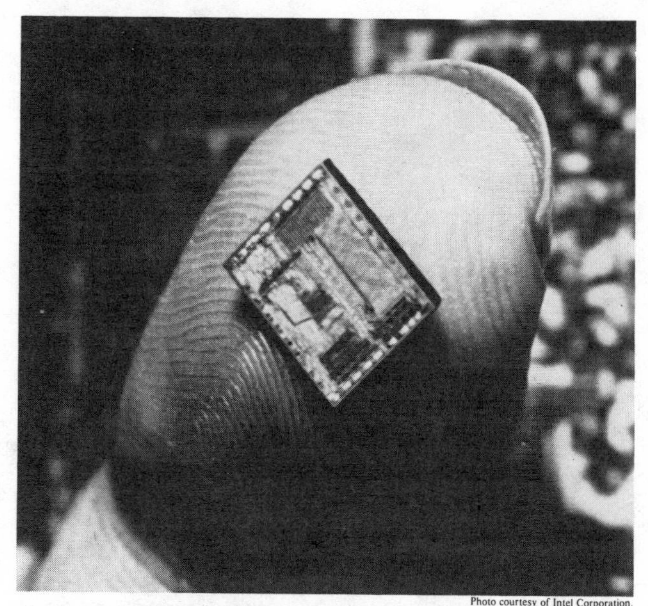

Photo courtesy of Intel Corporation.

A microprocessor chip fits on a fingertip and yet contains logic that once filled an entire room.

is a general purpose electronic circuit that can be used in calculators, watches, video games, or an infinite variety of other products, just as a computer can be programmed to perform an infinite variety of tasks.

Because the microprocessor is the basis of the microelectronics revolution, we will describe its origins.

In the "good old days" there were computers and there were computer terminals. Computers were big and expensive; a typical IBM computer of the late sixties could fill a small room and might cost $20,000 a month, or more, to rent. But that was just the beginning of the computer owner's expenses. The computer required a special air con-

Photo courtesy of University of California, Davis.

A typical computer room.

Photo courtesy of Hewlett-Packard.

Working at a computer terminal.

ditioned computer room, usually with a raised floor, beneath which cables could be run connecting different parts of the system.

The computer of the sixties was much like the steam engine of the railroad industry. It was big and costly to maintain. The economics of owning one was justified only by hundreds of passengers — or computer users. Passengers got seats in railroad carriages. Computer users got terminals.

But whatever a terminal might look like, its purpose is to give the computer user access to a large, expensive computer. A cable might connect the terminal to a computer in the next room, or the terminal might use telephone lines to communicate with a computer on the other side of the country.

Looking at a computer, enshrined in its cathedral-like computer room, you might assume that it is used to solve incredibly complex equations or to perform mysteriously difficult accounting tasks. Nothing could be further from the truth. Most computers spend the bulk of their time performing tasks that would not tax the learning of a child. But the computer is fast. It takes a computer one thousandth of a second to perform a task that might take a human hours.

Computer owners, like everyone else, tend to trade upwards. Having bought a million dollar computer, users quickly start eyeing the two million dollar version because the million dollar model has gotten overloaded. But it occurred to many people that if a million dollar computer spent most of its time performing trivial tasks, then perhaps money was being wasted. Instead of buying the two million dollar version, why not put a computer that is small, inexpensive, and simple right into the terminal? The terminal no longer serves merely to connect the computer and the user; it handles, in addition, many simple tasks itself. That way the million dollar computer ceases to be so busy, and the purchase of a two million dollar upgrade can be avoided!

An analogy for the large central computer and its many terminals is the old-fashioned family with too many children. The parents' home is equivalent to the large central computer. For a while, the entire family can live, eat, and sleep in the family home. When away from home, they can use a telephone to communicate, but the home provides for all of their needs. As more children arrive, the parents find that the home becomes too small. They could trade upwards for a bigger home; that is like trading upwards for the two million dollar computer. As an alternative, they could move the older children out into small apartments. These small apartments are equivalent to terminals with small computers inside them. The older children may still come home to do their laundry, store and retrieve belongings, or join the family gatherings, but they will sleep, eat, and entertain their friends in their small apartments. Similarly, someone using a terminal with a small computer inside it can use this small computer to handle routine necessities, relying on the big computer for major undertakings.

This analogy has more significance than is immediately obvious. The older child in an apartment is not necessarily dependent on the family home; using this as an illustration of dependence is a trifle contrived. The same logic is frequently true when looking at computers and their terminals — a fact that escaped most computer professionals, but that is the subject matter for Chapter 2.

In the late sixties, Viatron, Cogar, and Datapoint were three companies that attempted to exploit the idea of having computer terminals containing small computers. The fortunes of these three companies exemplify the risks and rewards facing anyone who wants to become an active participant in the microelectronics industrial revolution.

Viatron made financial history in 1970 by managing to spend thirty million dollars in one year, while generating just three million dollars of income. That spectacularly unimpressive performance resulted from Viatron attempting to build its own microelectronic devices. It also resulted in Viatron's bankruptcy.

Cogar succeeded in building a rather modest terminal containing its own tiny computer circuits. But by the time Cogar had its product working, it too was close to bankruptcy and had to be acquired. The buyer was Singer Business Systems, a division of the well-known sewing machine manufacturer. Unfortunately Singer's computers were almost as bad as its sewing machines used to be good. Acquiring Cogar simply added another hokey product to Singer's already hokey computer product line. Soon Singer's financial fortunes had declined to the point where the company fired its president and decided that it knew more about sewing machines than it did about computers. So Singer sold its computer business. And to whom? No American company would touch it, they knew better. In 1976, International Computers, Ltd. (ICL), a British company, was duped into picking up Singer's computer business. The total purchase price for the assets of Singer Business Machines that were acquired by ICL will be in accordance with an agreed formula based on audits now in progress. A down payment of 2.0 million U.S. dollars was made in April of 1976. A further $0.8 million was paid in November of 1976. The balance of the total purchase price will be paid in three equal installments in October of the years 1978, 1979, and 1980. A pitiful end to a massive undertaking.

Datapoint was smarter than Cogar and Viatron. It also wanted the entire computer on a single microelectronic circuit, but it contracted with Texas Instruments and Intel, two companies already established as leaders in microelectronics technology, to build the microelectronic circuit. Texas Instruments never came through; Intel did. But the product Intel produced, while doing everything Datapoint asked for, took more time in the doing than Datapoint had allowed. So Datapoint gave up on the idea of a single-circuit computer and manufactured its "smart" terminal (called the Datapoint 2200) using then conventional electronics.

That left Intel with a slow computer on a single microelectronic circuit, the designing of which had been paid for, but with no customers. Intel looked at the product, and said, "What the hell, we might as well try to sell it." And Intel did. All microelectronic devices are given numbers, not names. Intel gave its product the number "8008"; it advertised its 8008, and sat back waiting to see what would happen. Plenty happened. Customers seemed to crawl out of the woodwork, and the era of the microprocessor had arrived.

The 8008 is the grandfather of today's popular microprocessors. Intel sold its first 8008 in 1973. Reliable sales volumes for microprocessors are hard to come by, but an educated guess is that since 1973 Intel has sold more than three million 8008's and is still selling them.

2
The Fortunes of War

A computer terminal having a small computer inside it, as described in Chapter 1, is known as an intelligent terminal. The computer profession loves to make picturesque misuse of words when generating its own private jargon. "Intelligent" terminals are not intelligent; intelligence requires perception and understanding. Computers have neither. Nevertheless, the adjective intelligent has come to precede the name of any product that contains a microprocessor.

Computer professionals are frequently victims of their own misconceptions. Having put a small computer inside a terminal, they labeled the package an intelligent terminal in order to differentiate the big central computer, to which the terminal is connected, from the little computer inside the terminal. By implication, the big, central computer is a real computer, whereas the little computer inside the terminal is merely "local intelligence." But what if the little computer in the terminal were as powerful and as capable as the big, central computer, although a great deal smaller? No one believed this could happen. But it is happening, and whole new industries have emerged from this unlikely occurrence. These industries are known variously as the microcomputer or personal computing industries.

No event was more indicative of the revolution to come than the emergence of the microcomputer and personal computing industries. We can learn important lessons for the future from the manner in which these industries emerged.

The microcomputer and personal computing industries sprang out of nowhere, with products no one thought could exist, let alone be sold. Even the name is a misnomer. Personal computing conjures up the image of hobbyists, building computers at home for the sheer pleasure of it. Such people do exist, and in fact the industry had its origins catering to their needs, but today companies that started out building computers for hobbyists are generally building small business computer systems, doing a job established companies could have done but were not farsighted enough to anticipate.

The microcomputer and personal computing industries began as recently as 1974, when a company called Micro Instrumentation and Telemetry Systems (MITS), located in Albuquerque, New Mexico, designed and sold computer kits to be assembled by electronics hobbyists in their homes. At that time, MITS manufactured electronic calculators and scientific instruments; its financial health, however, had deteriorated to the point where bankruptcy was staring it in the face. Ed Roberts, one of the two principals at MITS, had long dreamed of building computers in kit form. He was convinced that, given a low enough price, there would be a substantial market for such a product. Facing bankruptcy anyway, he and his partner, Eddie Curry, decided to give kits a try. During the second half of 1974, they designed a computer kit around the "8080A", a microprocessor that had just been introduced by Intel. The 8080A was an upgrade of the 8008. Roberts' and Curry's banker demanded a business plan from MITS for this new, hare-brained scheme. So they worked up a plan that predicted 800 kits would be sold during 1975. MITS sent its first kit to Les Solomon of *Popular Electronics* magazine, and Les featured the kit on the front cover of the January 1975 issue. The results were stunning. Orders poured in, most with checks for the full amount of the kit. On one Friday afternoon shortly after *Popular Electronics* reached its readers, MITS received 400 orders — half of what it had planned to sell in a year. Eddie Curry thinks MITS shipped approximately two thousand kits in calendar year 1975 — all that it could manufacture. In May of 1977, two and one quarter years after shipping its first kit, MITS was acquired by Pertec for approximately six million dollars worth of Pertec stock. Pertec is a large company manufacturing a variety of computer-related products. In 1978 Pertec's gross revenues amounted to $131,802,000. Yet Pertec itself was founded as recently as 1967 by Stu Mabon, then a 29-year-old immigrant from England, an engineer with no prior business experience.

The second major manufacturer of personal

computers was a company known as IMS Associates. And its story is almost as extraordinary as the story of MITS. IMS Associates had been planning to develop a business computer system for General Motors automobile dealers. IMS Associates decided that the MITS kit would function admirably as the "computer" within the system. IMS Associates and MITS could not get together on prices and delivery terms, however, so IMS Associates decided to build its own microcomputer, patterned closely on the MITS kit. While shooting the breeze one day, Bruce Venatta, Joe Killian, and Bill Mallard, three principals at IMS Associates, decided there might be a side market for the microcomputer kit itself. So they ran some ads in *Popular Electronics* magazine. Like Roberts and Curry, they were overwhelmed by the response. In December of 1975 IMS Associates shipped 35 kits. In January of 1976 it shipped 400 kits. And from that point on, the number of kits it shipped was limited only by its ability to manufacture the product. Ultimately, IMS Associates assumed the more catchy name IMSAI, the name by which its products are known today.

Unfortunately, owing to poor management, IMSAI declared bankruptcy in March of 1979. It should have been the leader of the microcomputer industry.

The most interesting aspect of the IMSAI story is that IMSAI was one of the first MITS customers, yet IMSAI's was a purely business data processing application. Almost from the beginning, the major customer base for hobbyists' kits were business-oriented professionals, not hobbyists. In fact, by mid-1977 kits had all but disappeared, and microcomputer manufacturers were selling computer systems that were smaller and cheaper than, but otherwise identical to, products sold by established computer manufacturers.

An unfortunate aspect of the MITS and IMSAI stories was that these two companies usually required that their customers pay cash in advance with orders. This unsavory practice was a necessity, and it became an industry standard since banks are not equipped to deal with revolutions. Even today, banks are slow to fund the new crop of microcomputer manufacturers. In the absence of adequate financing, the multitude of companies that followed in MITS' and IMSAI's footsteps advertised products that did not exist and demanded cash up front with all orders. Then these companies spent the up front cash building the product. This dubious funding technique is referred to euphemistically as "forward financing." While it was frequently used with good intentions, it was also used dishonestly or, at best, questionably.

It is noteworthy that thousands of people were prepared to send hundreds or even thousands of dollars for products they had never seen to companies they had never heard of. These were not greedy speculators buying Florida swampland. They were otherwise intelligent human beings with an overwhelming desire to buy a computer once the price tag was low enough. This customer base was so large, and its appetite for the product so insatiable, that within four years almost two hundred companies came into being, catering to the need.

Among these new companies was Data Sync, a company that ran multi-page advertisements for nonexistent products for which customers had to

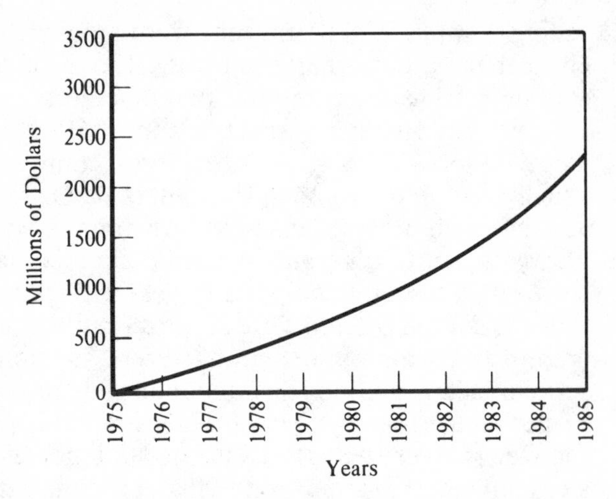

Author's estimates of past and future sales of the microcomputer and personal computing industries.

pay in advance. David Winthrop, the perpetrator of Data Sync, made off with a lot of money but was caught and sent to jail. He escaped early in 1979 and successfully repeated the same scheme, this time calling his company World Power Systems. Customers appear to have large appetites for such products and small memories for past scams.

Because the microcomputer and personal computing industries are so new and disorganized, no reliable estimates exist for their level of business. From my personal contacts with the industries, however, I have made my own estimates of past and future sales. While my estimate may seem large, it is worth noting that Radio Shack sold one hundred million dollars worth of microcomputer systems in the first 18 months that its product (known as TRS-80) was available, and Commodore sold perhaps half that dollar volume of PET's (its personal microcomputer system) in the same amount of time.

Can anyone explain how established computer manufacturers — and there were more than thirty of them in 1974 — missed such a large market? Why was this entire industry left to reckless entrepreneurs, lucky amateurs, and newcomers to computer manufacture? The answer is that this new market was too bizarre to fit any predictions made by established means. And over the next thirty years we will see similar scenarios — again, and again.

From early 1975 until mid-1977 almost anyone could have established a successful business selling microcomputer systems. And many did. But not one established computer manufacturer was significant among them.

Apple Computer Corporation was founded by Steve Jobs and Steve Wosniak when they were ages 24 and 20, respectively. They designed their microcomputer for the fun of it. Apple Computer Corporation will probably gross more than fifty million dollars in 1979.

Back in 1976, Bob Harp was working for Hughes Research and Development Company as an electronic engineer. His wife Lore and her friend Carole Ely were looking around for something to occupy their spare time. So Bob Harp designed a memory board that would fit into other people's microcomputers. A memory board is a relatively simple electronic module that holds the information used by a computer.

Bob Harp designed his memory board as a kit, which buyers had to assemble. In August of 1976, Lore and Carole set up production facilities in the bathroom of the Harp home, and proceeded

Photo courtesy of Vector Graphic, Inc.

The memory board designed by Bob Harp of Vector Graphic, Inc. Such a memory board, smaller than most hardcover books, holds all the information a computer would use.

to ship four thousand memory board kits during the next twelve months. Thus encouraged, they named their new company Vector Graphic, put together six thousand dollars of investment capital, and moved out of the Harp bathroom into more prestigious quarters. Vector Graphic now builds entire microcomputer systems. In 1979 it will probably gross six million dollars.

Then there is Chuck Grant who, back in 1976, opened a computer store with the whimsical name "Kentucky Fried Computers." At that point Chuck obviously wasn't taking his new business venture very seriously. The good Colonel sent Chuck a letter suggesting a name change. Chuck readily agreed, since he had started manufacturing

Photo courtesy of Vector Graphic, Inc.

Bob Harp

Photo courtesy of Vector Graphic, Inc.

Lore Harp, President, and Carole Ely, Secretary/Treasurer of Vector Graphic, Inc.

his own computer hardware and was doing a level of business that had exceeded his wildest expectations. His enterprise was no longer a game, it was a hugely successful operation. So Kentucky Fried Computers became North Star Computers and will probably gross ten million dollars during 1979.

Perhaps the most startling success story is that of Alpha Micro Systems. This company was founded on March 17, 1977, in Irvine, California, by Richard Wilcox, Robert Hitchcock, and John French. They put fifty thousand dollars worth of equipment and one thousand dollars cash into the new company. With no additional investment they sold approximately five million dollars worth of computer systems in 1978. They will probably sell at least twice that much in 1979.

Today, among the more than two hundred companies manufacturing products for the microcomputer marketplace, there are such well-known names as Texas Instruments, Radio Shack, and Commodore. Did you ever think you would see the day when IBM would compete with Radio Shack? Well, it does today. IBM is selling its smallest computer (the 5110) and Radio Shack is selling its TRS-80 in the same microcomputer marketplace.

It is entertaining to read about a bored housewife who makes a fortune assembling computer hardware in her bathroom, or about three computer professionals who turn a fifty thousand dollar investment into a major computer manufacturing operation within two years. But while all this was occurring, where were Digital Equipment Corporation, Data General, and Hewlett-Packard, the three leading minicomputer manufacturers? Where were the other thirty-odd computer

Photo courtesy of Radio Shack, a division of Tandy Corporation.

Radio Shack TRS-80 "Business" System competes with IBM.

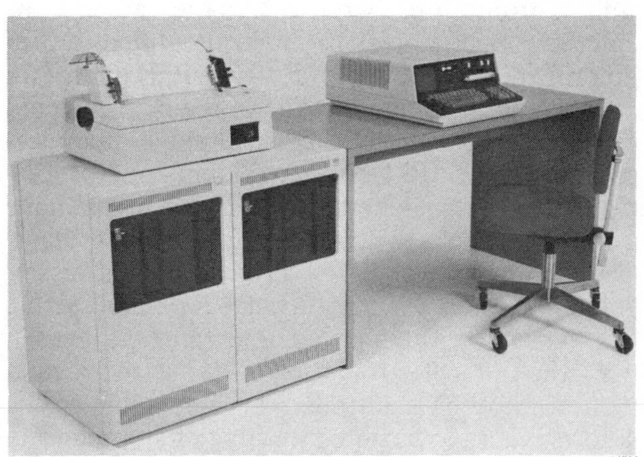

Photo courtesy of IBM.

IBM 5110

manufacturers? Were they so big that they could ignore an imminent billion dollar market? Indeed no. Their managements simply could not understand what was going on around them. And if these industry professionals could not understand, what chance does a layman have?

The microcomputer story is not unique. Over

the next twenty-five years, the industrial revolution will replay the same story many times, with more new industries springing up under the devastated noses of those who should have seen them coming.

Consider the helpless position of any special interest group trying to stop this new industrial revolution. Take the hypothetical case of a union that evaluated microprocessors back in 1975 and was farsighted enough to realize the long range threat of microprocessor-based automation to union jobs. Assuming that this union had had the wit to foresee that the first step, preceding robots and automation, would be the advent of microcomputers, it might have attempted to control the microcomputer industry as it formed. And how would the union have gone about it? It would have attacked established computer manufacturers such as IBM, Digital Equipment Corporation, and Data General. Would the union have gone after a housewife assembling computer hardware in her bathroom? Or a couple of kids designing a computer for the fun of it? Obviously, the answer is no. And that demonstrates one reason why the microelectronics-based industrial revolution cannot be controlled or stopped: no one knows where to begin controlling or stopping any aspect of it.

3
Computer Intelligence?

The previous chapters argue that the microelectronics revolution can neither be controlled nor stopped since no one can tell what the next development will be or where it will occur.

But how dangerous is this inability to control? That surely depends on the ultimate capabilities of microelectronics. And in this context nothing is more significant than the question: is electronic intelligence capable of duplicating human intelligence?

At this point in time, no one can state definitively whether a computer will ever rival the human brain. In some ways the computer is clearly superior to the human brain. For example, no human, even the rare mathematical genius, can add with the error free speed of a computer. A very inexpensive computer can add any two four-digit numbers (in the range 0 through 9999) within ten millionths of a second. In other words, the computer can perform one hundred thousand such additions in one second.

The most expensive and powerful computers available in 1979 can perform one hundred million such additions in a single second. But computers do only what they are explicitly told to do, whereas the human brain is capable of "constructive thought." Confronted with a new situation, the human brain will invent solutions. Solutions may be derived by drawing on analogies from previous experience or by making seemingly unrelated juxtapositions. If this defines constructive thought, then computers are not capable of constructive thought — today. But what about tomorrow? If computers were capable of thinking in the human sense of the word, then the combination of an ability to think and the computer's operating speed presents the unsettling prospect of a machine that is intellectually far ahead of man.

I do not believe that electronic logic will ever rival the human brain — but I cannot prove my case. However, no one who believes electronics will one day rival the human brain can prove his or her case either. Therefore, in this chapter I will explore both sides of the argument.

Superficially, the most striking difference between electronic and human intelligence is this: the human mind can be inventive, whereas a computer will do only what it has been explicitly instructed to do, nothing more and nothing less. We refer to a computer's explicit instructions as a computer program; the people who create computer programs are thus called programmers.

To illustrate the fundamental difference between computer and human intelligence, we will look at two examples, both of which have enlightening, though not immediately obvious, ramifications.

Consider first the simple example of addition. A person learns the concept of addition and from that point on attempts to add correctly but is not always successful. In other words, the human brain understands what it is supposed to do, even though (with the best of intentions) its answer is not always correct. A computer, on the other hand, is given a set of explicit instructions that cause two numbers to be added, accurately, in consequence of some fixed logic sequence.

Consider the addition of two single digit numbers. We can construct a table of answers.

		Augend									
		0	**1**	**2**	**3**	**4**	**5**	**6**	**7**	**8**	**9**
Addend	**0**	0	1	2	3	4	5	6	7	8	9
	1	1	2	3	4	5	6	7	8	9	10
	2	2	3	4	5	6	7	8	9	10	11
	3	3	4	5	6	7	8	9	10	11	12
	4	4	5	6	7	8	9	10	11	12	13
	5	5	6	7	8	9	10	11	12	13	14
	6	6	7	8	9	10	11	12	13	14	15
	7	7	8	9	10	11	12	13	14	15	16
	8	8	9	10	11	12	13	14	15	16	17
	9	9	10	11	12	13	14	15	16	17	18

The number stored where the augend column and addend row intersect is the sum of the augend and the addend.

Table of addends and augends and their sums.

A person will quickly understand concepts embodied in the table above and will then discard the table. But electronic logic will never gain such conceptual understanding. A computer's addition program must, instead, rely on some rote logic,

Addend		Augend									
		030	031	032	033	034	035	036	037	038	039
		20	21	22	23	24	25	26	27	28	29
040	00	0020	0021	0022	0023	0024	0025	0026	0027	0028	0029
		0	1	2	3	4	5	6	7	8	9
041	01	0120	0121	0122	0123	0124	0125	0126	0127	0128	0129
		1	2	3	4	5	6	7	8	9	10
042	02	0220	0221	0222	0223	0224	0225	0226	0227	0228	0229
		2	3	4	5	6	7	8	9	10	11
043	03	0320	0321	0322	0323	0324	0325	0326	0327	0328	0329
		3	4	5	6	7	8	9	10	11	12
044	04	0420	0421	0422	0423	0424	0425	0426	0427	0428	0429
		4	5	6	7	8	9	10	11	12	13
045	05	0520	0521	0522	0523	0524	0525	0526	0527	0528	0529
		5	6	7	8	9	10	11	12	13	14
046	06	0620	0621	0622	0623	0624	0625	0626	0627	0628	0629
		6	7	8	9	10	11	12	13	14	15
047	07	0720	0721	0722	0723	0724	0725	0726	0727	0728	0729
		7	8	9	10	11	12	13	14	15	16
048	08	0820	0821	0822	0823	0824	0825	0826	0827	0828	0829
		8	9	10	11	12	13	14	15	16	17
049	09	0920	0921	0922	0923	0924	0925	0926	0927	0928	0929
		9	10	11	12	13	14	15	16	17	18

Address (shaded) **Content** (unshaded)

Addition table in which the addends, augends, and sums have been arranged so that each number is held in an addressed location as it would be in computer memory.

represented as a sequence of instructions. One possible sequence of instructions might depend on all numbers shown in the table being held in addressable computer memory locations, which you might visualize as numbered pigeon holes.

(You should place no special significance on the addresses that have been selected for the illustration. Actual numbers are used simply to make the illustration easier to follow. Since addresses range as high as 929, the computer memory must have at least that many addressable locations, each capable of holding a complete number.)

The ten addressable memory locations reserved for the addends contain numbers 0 through 9 — as you might expect — but the ten augend addressable memory locations contain values 20 through 29. These are not the required augend values. However, the logic we are putting together does not require that either the augend or the addend contain actual augend and addend values within their respective memory locations; the fact that the addend contents are real addend values is coincidental. The key to the logic is the fact that the addend memory location contents together with the augend memory location contents create the memory address within which the sum will be found. Suppose you must add five and seven. Five is the sixth addend value since values begin at zero. The sixth addend memory location has address 045. In this location, you will find 05. Similarly, seven is the eighth augend value, since augend values begin at zero. The eighth augend memory location has the address 037, and in this location we find the number 27. Concatenating the addressed addend and augend memory location contents, we create the address 0527. You will find stored in this location the value twelve, which is the sum of five and seven.

In order to use the illustrated table, logic for a computer's addition program might appear as follows:

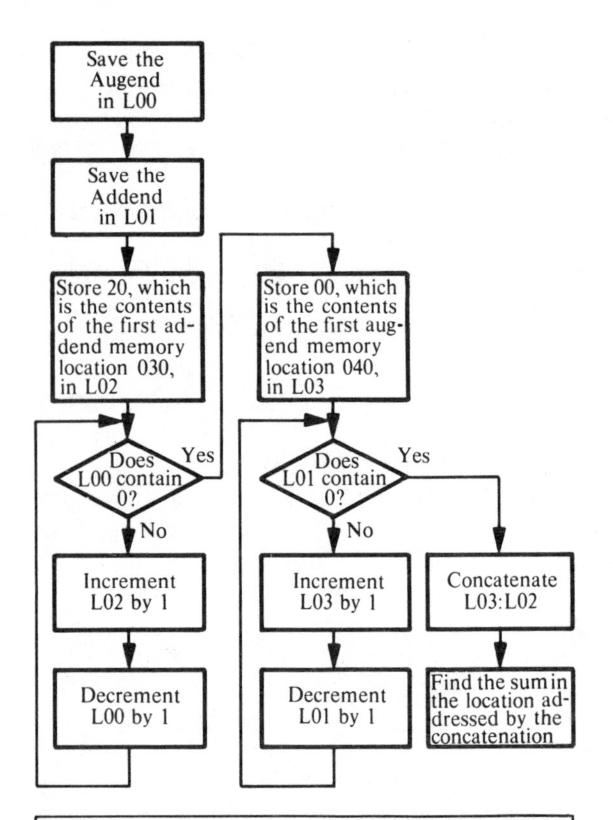

L00, L01, L02 and L03 are four unused storage locations which we are going to use while adding. "L*xx*" is shorthand for "addressable memory location whose address is *xx*"

If you blindly follow the computer's addition program logic sequence in conjunction with the numbers table as illustrated, you will always obtain the correct sum of two digits even if you do not understand the concept of addition.

What if one of the numbers in the table is wrong?

		Augend									
Addend		**030**	**031**	**032**	**033**	**034**	**035**	**036**	**037**	**038**	**039**
		20	**21**	**22**	**23**	**42**	**25**	**26**	**27**	**28**	**29**
040	**00**	0020	0021	0022	0023	0024	0025	0026	0027	0028	0029
		0	1	2	3	4	5	6	7	8	9
041	**01**	0120	0121	0122	0123	0124	0125	0126	0127	0128	0129
		1	2	3	4	5	6	7	8	9	10

The computer will consistently generate incorrect answers whenever the augend is five if, as illustrated in the table above, the number stored in memory location 034 has had its digits switched; the number 42, not 24, is present.

What if the instruction sequence itself is wrong? Consider the following possibility:

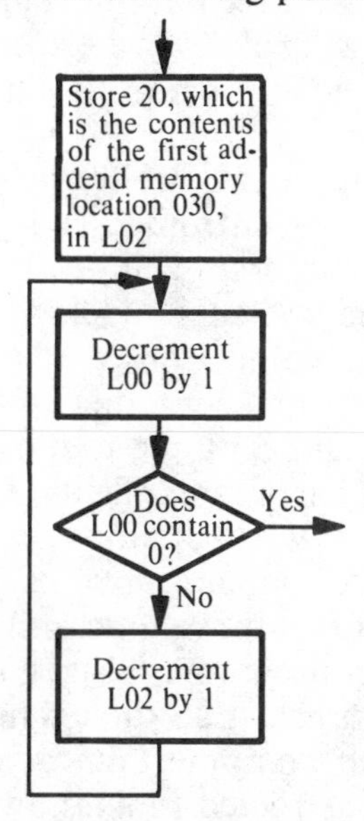

We are using logic that assumes that augend numbers begin at zero instead of one. In consequence, the computer will generate the sum of the augend, plus the addend, plus one. It is, after all, blindly following instructions. A person might spot inconsistencies or obvious errors because the human brain understands what addition is all about. The computer does not understand the concept of addition; it does exactly what its program specifies, however absurd the consequences.

Consider a more graphic example. Computers have been programmed to play chess. In fact, the best chess playing computer programs can beat any human, short of a grand master. Nevertheless, computers do not "think" while they play chess as people do. They respond to a program — an absolutely and explicitly defined set of rules that leave nothing to judgement. These instructions will be vastly more complex than, but conceptually identical to, the illustrated addition logic.

But suppose we arbitrarily change the way knights move. Confronted with this new situation, the human brain will invent solutions. Relying on past experience of playing normal chess, people will have no trouble adapting to the new knight move. But a computer could not cope. The computer would continue to move the knight the old way until it was explicitly reprogrammed to specify the new knight move. And then it will specify the new knight move using strategy based on the old knight move unless the strategy part of the computer program is also changed. Moreover, the smallest mistake in the changed program will result in the computer doing whatever it was mistakenly instructed to do, however absurd.

The human chess player has learned how to think. The computer cannot learn to think. It

This is how knights are supposed to move.

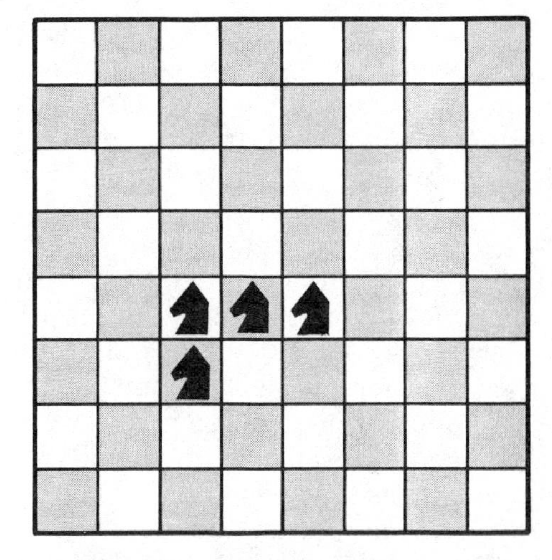

This is our new move.

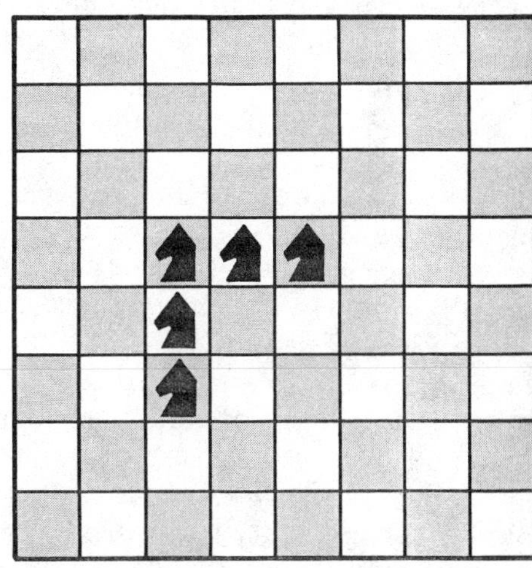

must be given an explicit set of instructions that leaves nothing undefined.

There are two ways of looking at the two examples illustrating the difference between electronic and human intelligence. First, we can use these two examples as a simple illustration of how people can think, while electronic intelligence cannot. But it is equally valid to claim that the examples are flawed because they make unwarranted assumptions regarding the nature of human intelligence.

What is the human ability to think? Should we begin all computer programs with a program that allows the computer to think like a human? What then? If you look at the addition example, electronic logic is allowed to make dumb mistakes no human would make because electronic logic has been denied the ability to understand. Is this a fair comparison? What if someone could indeed write a computer program that gave a computer every appearance of "understanding"?

That would be some undertaking.

Defining exactly how the human mind thinks, so that we could program a computer to think, would be a formidable task. The problem is that electronic logic begins at a very elementary level of capability. That is, a computer program must define a problem explicitly for the computer to solve that problem because electronic intelligence is so primitive.

The human thought process is far from primitive. But perhaps the human thought process is the consequence of a very complex program depending on some biological logic that is just as primitive as elementary electronic logic. If this is the case, then perhaps someone can write a computer program that would give electronic logic the capabilities of human thought and understanding.

Carl Sagan, in his book *The Dragons of Eden**,

*Carl Sagan, *The Dragons of Eden*, Ballantine Books, New York, 1978, pg. 47.

estimates that the logic in an average human brain is equivalent to ten thousand billion (10^{13}) elementary electronic logic circuits. If Carl Sagan's estimate is accurate, then the amount of logic is not an overwhelming problem. Within twenty-five years you will likely be able to buy enough electronic logic to match the human brain for much less than one million dollars, perhaps for a few thousand dollars. Assuming that all other problems could be solved, this price tag would keep the thinking computer outside of the average home, but it would certainly be economically viable as a commercial product. Large corporations have for years been spending such sums for data processing computers.

But can all other problems be solved? I believe not. We are a very long way from having the most rudimentary understanding of how the human brain works — let alone having the utterly complete and totally accurate understanding needed in order to create an electronic equivalent. If we can completely understand, down to the smallest detail, how the brain thinks, then the electronic equivalent of human intelligence may be possible. But even then, we can only say maybe, since there is no guarantee that electronic logic will be able to duplicate the patterns of thought once they are understood.

4

The Blue Collar Robot

Electronic technology has created new industries and jobs; it has also meant the wholesale elimination of selected crafts and professions. Consider phototypesetting. Until the early seventies, all newspapers and books had their type set using special Linotype machines that were manned by Linotype operators. Today phototypesetters connect directly to editing computer terminals that are operated by typists and reporters, and the Linotype operator's profession has gone the way of the blacksmith's. On the other hand, electronics created the television industry — and all the jobs that go with it.

Technology has been creating and eliminating jobs for the past two hundred years, ever since James Watt invented the steam engine. Do unions and the blue collar work force have any special reason to fear microelectronics?

Indeed they do.

Because microelectronics, for the very first time, has allowed machines to compete with people for jobs that depend on the intelligence of the human brain.

Chapter 3 concludes that electronic intelligence is unlikely to rival the human brain, at least in the immediate future. But electronic intelligence does not need to rival the human brain before it starts replacing people. Many jobs make limited use of human intelligence, enough to have kept machines out in the past, but no longer.

Microelectronics will begin by replacing jobs that do not require too many "smarts." Supposedly these are blue collar jobs, although some might argue the point. The last jobs to go will be those that depend on human judgement and opinion.

To describe the possible impact of electronics on every job and industry would be a hopelessly ambitious task. At the blue collar level, therefore, this chapter will examine one product, the industrial robot, and one industry, the mail service. By analogy you can then decide for yourself what impact, if any, electronics will have on any other industry or job.

First consider robots.

The word robot conjures up the image of a thinking, metallic, human-like figure. In fact, if you saw an industrial robot, you would probably not recognize it for what it is. The industrial robot manufactured by a company called Unimation, for example, is a single "arm," with one set of "pincers."

Industrial robots of the Unimation type were first manufactured in the late sixties. They were very expensive because at that time the electronic logic needed to support a mechanical arm was costly and complex. In consequence, robots were used only where people could not work: where it was too hot, in the presence of a radiation hazard, or on the ocean floor. For example, following the Three Mile Island, Pennsylvania, nuclear power plant accident (March 1979), a robot was used to take samples of radioactive water inside the damaged reactor.

The one-armed robot is a conceptually simple device. Six electric motors can control every movement for a limb with two joints, a rotating wrist, and one pair of pincers. Such a robot is programmed by specifying which electric motors are to operate, when, and for how long.

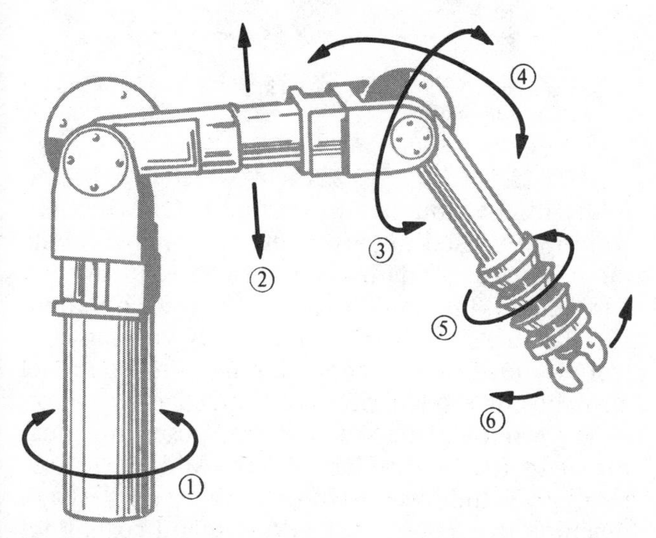

Every movement of a Unimation type robot can be driven by six electric motors. Motors 1 and 2 rotate and angle the arm. Motors 3 and 4 move the "elbow" up and down and from side to side. Motor 5 rotates the pincers. Motor 6 opens and closes the pincers.

Mobile robots are available today. They roll on wheels rather than walking on legs. A wheeled robot can negotiate neither stairs nor obstacles, but it is easier to build a flat surface for a wheeled robot than to build a legged robot capable of negotiating an obstacle course.

A robot tracking a wire: a method used in warehouses to help robots move along a fixed path.

There are many ways in which a wheeled robot might find its way around. One method commonly used in warehouses, where robots move along fixed paths, is to lay a current-carrying wire down the middle of the path. The robot simply tracks the wire. Alternatively you can control a robot's movements via radio, just as model airplanes or model boats are controlled by radio.

A simple, one-armed robot poses no great threat to the blue collar worker. Add an arm or two, give the robot wheels; you still have a machine that is not very versatile and costs a lot for what it does. But what if robots were smart and inexpensive? Certainly they have better work habits than people.

Robots work without getting tired, they cannot be distracted, and they do not make mistakes. Humans get tired and distracted; they also make mistakes.

A robot can work continuously, twenty-four hours a day, seven days a week, if necessary. People like to eat, sleep, play, and take coffee breaks.

Robots do not belong to unions, and they do not file job grievances. People do.

What, then, can electronics do for robots? Let us look at some of the problems facing robots today and see how these problems might be solved.

A mechanical arm that can make exactly specified movements sounds useful. But it has problems. What if the base of the robot is jogged or shifted? What if the arm must pick up an object on a moving belt or place the object on a moving surface? Having an arm that moves through an exact path no longer works, because the robot will not be so lucky as to find the object correctly positioned every time.

There are a variety of ways in which robots can be designed to identify moving objects. We might put a light source on a moving belt or reflect light off white paint spots. A light cell no different than that found in most cameras can identify the light source, and electronic intelligence can use this information to control the robot arm, causing it to pick up or put down an object some exact distance from the light source.

In the illustration, electronic intelligence causes the arm to pick up a bolt six centimeters in front of, and two centimeters to the side of, a white spot on a moving belt. The robot arm can now cope with moving objects — a small step in

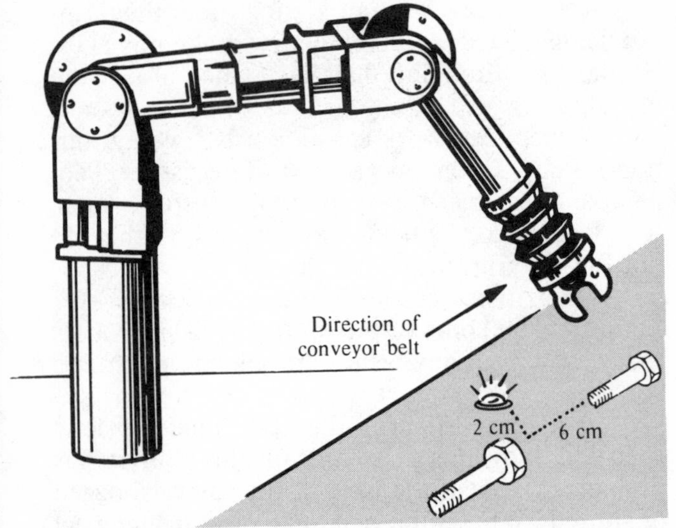

One of the possible ways robots might locate moving objects exactly. A light cell like the one in most cameras could identify light sources on a moving belt for the electronic intelligence controlling the robot arm.

the direction of intelligence.

The object itself might carry a marker. For example, a dark bolt (or any other object) moving on a dark conveyor belt could be identified by spots of white or fluorescent paint. The robot arm will now home-in on the light reflected by the paint spot. The robot arm has gotten smarter; it can pick up an object that is rolling around on a moving belt.

But the ultimate robot will be able to see. Fairchild Camera and Instrument Corporation and Texas Instruments have both designed microelectronic circuits that "see" by dividing up a field of vision into squares of varying light intensity.

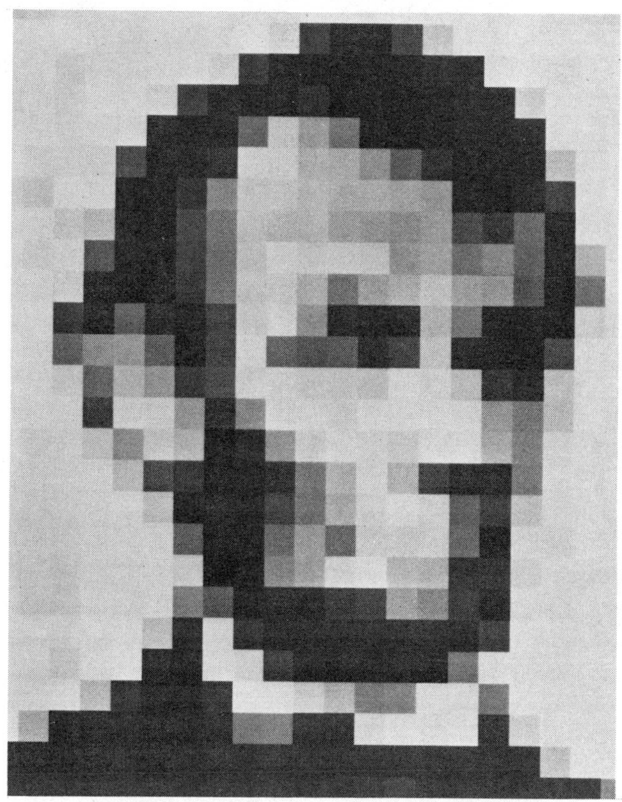

Photo courtesy of Blocpix, a division of Watson-Manning.

Spatially quantized image of Lincoln as "seen" by a computer.

In 1979 microelectronics that "see" in this manner are expensive. Therefore, they are not in common use. But like all things microelectronic, it would become inexpensive the moment a large market for it existed, and robots certainly promise a large market.

A robot with "seeing" electronics can identify an object or a location by its shape. Viewing objects on a moving belt, a robot might spot-weld

Photo courtesy of Texas Instruments, Incorporated.

Texas Instruments' "seeing" robots at work.

two pieces of metal, identifying the spot-weld point by shape. Texas Instruments already has robots that "see", working on calculator production lines.

Electronics can already hear crudely, too, and speak well.

We create electronic speech by recording word sounds, then stringing these word sounds together to generate whole words, sentences, or messages. This requires about as much electronic technology as is in a tape deck.

Electronic speech is already so inexpensive that Texas Instruments packs it into a $50 educational toy called "Speak and Spell". This toy

Photo courtesy of Texas Instruments, Incorporated.

Texas Instrument's Speak and Spell toy. Electronic speech, requiring about as much technology as a tape deck, is already so inexpensive that, with the small fifty-dollar toy, children can learn and practice their spelling while they play.

speaks words and asks you to spell them, using a keyboard. The toy then tells you if your spelling is right or wrong. (The March 1979 newsletter of the Amateur Computer group of New Jersey explained the secrets of this toy, showing how anyone could get inside it and teach it new tricks.)

"Hearing" electronics converts sound waves into number sequences. Electronic intelligence attempts to discriminate between sounds by comparing new sound wave number sequences with a stored library of old sound wave number sequences.

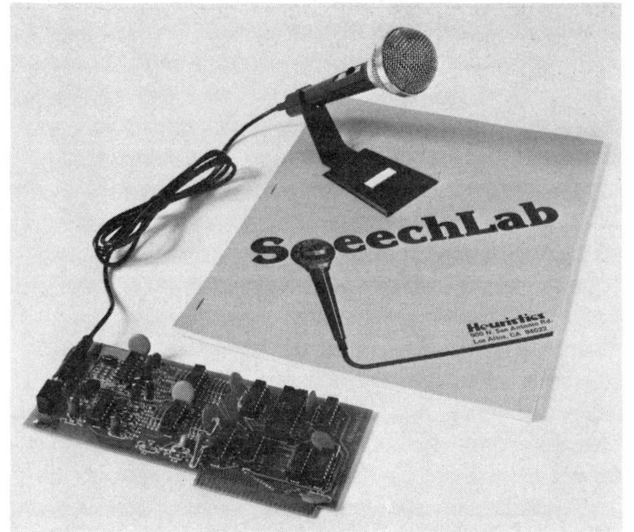

Photo courtesy of Heuristics, Incorporated.

A Heuristics® board able to identify 64 words, provided the person who recorded the words also speaks them.

"Hearing" electronics must, of necessity, be connected to a microphone.

First you set the electronics to record, and you speak the words you want recognized, one at a time, into the microphone. The "hearing" electronics converts the word sound wave patterns into sequences of numbers, which are stored in electronic memory as a library of words.

Next you set the electronics to "hear".

To hear words electronically, a device must compare the sound wave pattern generated by someone's speech with the stored library of such sound wave patterns. This is no simple task. The same word spoken by individuals with different accents can differ more than two words spoken by the same individual. For example, a computer may have to recognize "shedyl" and "skedjool" as different pronunciations of the same word "schedule"; "Iy" may be "I", or "eye", or perhaps a badly pronounced "aye."

In 1979 you can buy inexpensive electronics that hear, but its hearing is not very accurate. For example, a small company called Heuristics will sell you, for less than $200, electronics that can identify sixteen different words. Another Heuristics board is available which can identify up to sixty-four words.

Inexpensive hearing electronics, such as that manufactured by Heuristics, will get words right nine times out of ten, provided the same person who recorded the words speaks them. But if this individual catches a cold, or gets new false teeth, or if someone else speaks the words, recognition accuracy may go down sharply.

Researchers at Bell Telephone Company's central research laboratory claim to have designed electronic hearing devices that can decode anyone's speech, taking accent into account. They believe a general purpose electronic hearing system with a vocabulary of a few hundred words could be built today, to sell for less than $300. They believe such electronic hearing systems will be generally available within three years.

Thus the robot of the future will be able to pick up and move objects, operate machines, insert, weld, and manipulate, provided its mechanical limbs are dexterous and strong enough for the task. No task will be too difficult for the robot.

The robot of the future will be able to hear spoken words and non-verbal sounds and will be able to respond to what it hears. The robot of the future will be able to speak with an extensive vocabulary and generate sounds of almost any type. The robot of the future will be able to combine what it sees and hears with the movements of its limbs to perform any task that can be explicitly defined.

Given the tasks that robots will be able to perform, their impact on the blue collar force will be profound. Most assembly line jobs will be eliminated. Automobiles, washing machines, and television sets will all be assembled by robots. Robots will even assemble themselves.

Some twenty percent of today's workers have assembly line jobs. Ninety percent of these jobs could be eliminated over the next twenty years. But will they? Will unions stand still while they watch the wholesale elimination of their membership base?

If an industry does not face international competition, then special interest groups can deflect industry practices from an economic optimum. For example, the existence of railroads in any country outside of the U.S.A. poses no competitive threat to the United States railroad industry. You can haul neither cars nor passengers between Detroit and New York using the Japanese railroad system. Thus railroads in the U.S.A. can apply whatever pressures they choose to keep labor saving changes out of the United States railroad system. So long as some competitive transportation system (such as trucking) does not put railroads completely out of the business, inefficiencies introduced by pressure groups can be tolerated. But industries that face direct competition from other countries can afford no such luxuries. Consider the United States automobile industry. If no automobiles could be imported, then

UNITED KINGDOM
MOTOR CYCLE INDUSTRY
Includes mopeds, scooters and motor cycles

Production quantities given in thousands

1959	1969	1978
234.3	71.2	27.2

Source: The Motor Cycle Association of Great Britain Ltd.

the domestic industry could keep raising its prices until owning a car, despite its conveniences, became too expensive as compared to using public transportation. Labor unions could lobby to keep labor saving and cost cutting practices out of the automobile industry until automobiles themselves became uneconomical. But so long as Germany, Japan, and other countries are exporting automobiles to the U.S.A., the domestic industry must remain competitive with imports, or it will not survive. Therefore, if labor unions are to keep robots off production lines, they must create barriers against imports, or they must keep robots off production lines in every country because the one country they miss will put the others out of business. The British motorbike industry is an excellent case in point.

In one decade the British motorbike industry went from world domination to insignificance. Many have argued that it was the constant warfare between labor and management that put this segment of British industry into oblivion.

Within twenty-five years robots will be assembling automobiles, and electronic intelligence will be handling any manufacturing function that can be explicitly defined.

Microelectronics will help some existing industries and hurt others. For example, robots might reduce labor costs and therefore reduce prices.

If microelectronics could eliminate half or three quarters of the labor costs involved in manufacturing televisions, cars, washing machines, and clothing, the resulting decline in product prices could significantly increase sales levels.

But there are industries microelectronics can only cripple or eliminate.

Consider the post office and the mail service. Many might argue that the U. S. Postal Service has crippled itself so thoroughly that electronics has nothing left to do. Nevertheless, without new microelectronic innovations, the postal service could last forever, with all its inefficiencies. But because of new microelectronic innovations, the post office as we know it today will disappear within the next twenty years — possibly within the next ten years.

Already there are office machines on the market that transmit over telephone lines anything black and white that you can put on a piece of paper. One such machine is the QWIP®, manufactured by a division of Exxon Enterprises.

Today it costs between fifty cents and one dollar to send one page via QWIP between any two points in the U.S.A. The post office can increase first class postage quite a bit before QWIP Systems offers a serious economic threat. But as the cost of mailing a letter increases inexorably, while the cost of transmitting information via QWIP is likely to decrease just as inexorably, the

RECORD OF EMPLOYMENT COSTS BY INDUSTRY GROUPS

Based on S&P Annual Surveys

Salaries & Wages as % of Net Sales

	1973	1974	1975	R1976*	1977
Aerospace & Aircraft	37	37	35	34.9	36.2
Air Transport	40	36	40	38.4	39.8
Apparel	...	...	28	25.1	23.3
Autos & Parts	32	33	30	30.0	20.4
Beverages	24	15	16	18.2	18.2
Building Materials	33	32	28	27.4	27.0
Chemicals	25	22	25	24.7	24.3
Communication (Publishing, Broadcasting, Adv.)	...	...	39	39.3	38.8
Containers	36	33	33	31.2	31.6
Electrical-Electronics	38	37	36	35.3	35.1
Food Processing	19	17	16	16.5	16.8
Health Care & Cosmetics	27	27	28	27.7	27.8
Home Furnishings	33	29	31	31.3	31.1
Leisure-Time	37	33	39	38.9	38.0
Machinery (incl. Pollution Controls & Rail Equipment)	...	...	32	30.8	30.9
Metals—Nonferrous	33	29	34	31.7	31.8
Office Equipment	37	37	36	41.0	40.5
Oil	9	6	7	6.7	6.7
Oil-Gas Drill. & Services	32	32	30	30.7	31.3
Paper	28	25	26	26.8	27.3
Printing & Publishing	32	30	...	...	...
Retail Trade	16	16	15	15.3	14.4
Rubber Fabricating	32	30	30		
Steel	36	31	36	36.7	36.7
Textiles	...	...	30	29.2	29.1
Trucking	...	57	59	54.2	51.6
Average	25.6	22.8	23.9	23.8	24.0

*Data prior to 1976 not strictly comparable because of changes within groups. R-Restated.

Source: Standard and Poor's Industry Surveys, October 12, 1978 (Section 4)

crossover point is not far off.

Today QWIP machines transmit and receive in black and white only. Late in 1979 QWIP units will be available with a limited ability to handle color. Within n years QWIP-type machines will transmit and receive in full color, with very high reproduction quality. And that has implications for greeting card manufacturers. Because in ten years, instead of buying greeting cards, you might just sit in front of your television set examining pictures displayed on your screen until you find the one you like. You could then transmit the selected picture, together with an appropriate handwritten message, via a full-color QWIP transmitter/receiver. The card, with message, will be created by a QWIP at the receiving end.

The postal service will still have to handle packages and non-printed material. But the post office has plenty of competition in these areas.

Transmitting mail over telephone lines brings together the postal and telephone industries.

There are some interesting alternatives to today's QWIP unit. Instead of running a piece of paper through a QWIP terminal, how about creating information at a television screen, as proposed above for greeting cards? After all, a television is equivalent to the display portion of a computer terminal. You could type a letter at home, look at what you type on the television screen, and transmit this information over telephone lines to be received at the other end by another television.

Matsushita, the Japanese television manufacturer, is already selling television sets with built-in printers that will give you a full color print of anything displayed on the television screen.

Photo courtesy of Qwip Systems, a Division of Exxon Enterprises, Inc.

Exxon's QWIP Two allows the user to send correspondence and any graphic material anywhere in the world in three minutes. This kind of low-cost, high-speed unit will soon render obsolete the post office as we know it today.

Most telephone messages today are broadcast through the air in a manner analogous to radio or television. Telephone messages are no longer sent through cables. So why not take the next small step and transmit text through the air waves? It is already happening; the postal, telephone, and broadcasting industries find themselves in a new, curiously intermingled service.

In the United Kingdom, the postal and telephone services are jointly run by the general post office — a branch of the British Government. Broadcasting is largely government controlled. The British Post Office was among the first to see the advantages of electronic mail and text broadcasting. In consequence, Britain became a leader in this area, developing "Teletext," "Viewdata," and "Presetel" systems.

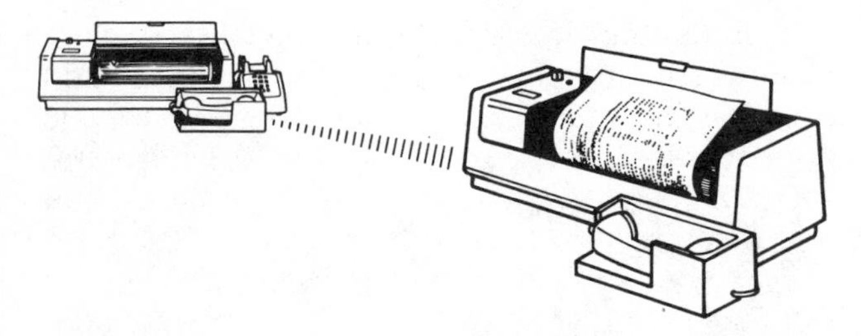

Transmitting mail over telephone lines brings together the postal and telephone industries.

From TV to TV a picture is sent by cable.

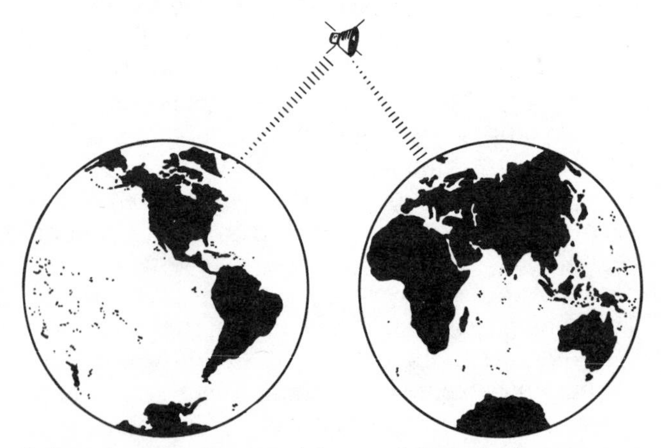

Perhaps the next step–transmitting text through the air waves.

But in the United States the postal service is a government agency, while the telephone and broadcasting industries are separate and private entities. Moreover, various United States government agencies regulate the U.S. telephone and broadcasting industries, and much of the governmental regulation is beset by the conflict between personal privacy and freedom of speech. Therefore, in the United States the post office has not had the freedom of its British counterpart to experiment with electronic mail or its many ramifications. Indeed, the future for the United States Postal Service is singularly bleak because a variety of large corporations could, and probably will, shut it out of the electronic mail industry. Companies such as American Telephone and Telegraph (AT&T), International Telephone and Telegraph (ITT), International Business Machines (IBM), and Xerox Corporation are vigorously pursuing all forms of electronic information transmission. Politicians are unlikely to legislate a monopoly for the United States Postal Service, and without the protection of a monopoly, the United States Postal Service is unlikely to survive.

How about your job? Or your industry?

That part of your job that can be exactly defined, and requires no human judgement, can be replaced by electronic logic. And if it can be, it probably will be. That part of your job that really does depend on human judgement is safe.

5
The White Collar Future

Microelectronics will confront the blue collar worker in a variety of strange disguises. But in white collar work, microelectronics will usually show up in a computer, a computer terminal, or a computer disguised as a business machine. Any discussion of the impact that microelectronics will have on white collar jobs must therefore begin with a look at the future of computers themselves.

IBM has been sued variously by competitors and by the United States Government for monopolizing the computer industry. Were the United States Government to win its case, IBM could well be forced to separate itself into a number of smaller, competing companies. But it will take at least five years for the U.S. Government suit against IBM to reach a final verdict. By then IBM will no longer have a case to defend; it will no longer be a significant supplier of computer systems. That is not to suggest that IBM is heading for bankruptcy and oblivion — far from it. IBM will continue to do an excellent business selling very large computer systems in an ever-dwindling market for such giants. But IBM's main business will be data communications, in direct competition with ITT, AT&T, and Xerox Corporation. IBM's general office products division will also remain strong. But in its most famous business, run-of-the-mill business computer systems, IBM will soon cease to be a significant force. Even while the government is suing IBM for being a computer monopoly, literally hundreds of new companies are bursting into existence, building very small computer systems today and bigger ones tomorrow. Soon IBM will be facing stiff competition, not from its traditional rivals such as Burroughs, Univac, Honeywell, and NCR, but from such upstarts as Apple Computer Corporation, Radio Shack, Commodore, Pertec, Texas Instruments, and the whole wolf pack of little guys. And when the IBM computer salesperson shows up at an office trying to interest the office manager in a new computer, that office manager may well be looking over a variety of computer systems at a local computer store.

The world's first computer store was probably "The Computer Store," which Dick Heiser opened in Los Angeles back in 1975. By the end of 1978 there were approximately seven hundred computer stores in the U.S.A. alone, and their numbers are growing rapidly. This explosive growth is clear evidence that computer stores are economically viable. Recently, Digital Equipment Corporation, the world's leading minicomputer manufacturer, started experimenting with com-

Photo courtesy of Dick Heiser.

Dick Heiser's "The Computer Store" opened in 1975 in Los Angeles, making it in all probability the world's first computer store. A short three years later, 700 such stores had opened their doors in the U.S.A.

puter stores. But with this timid exception, not one established computer manufacturer, not a single electronic parts distributor, nor any computer marketing organization opened a computer store. Why? The answer is that they never saw it coming. Yet computer stores are economically very viable for two simple reasons:

1) They give you the chance to examine a variety of different systems before you buy.
2) They eliminate sales representatives, who typically cost forty percent of what you pay for a computer system.

Many computer professionals and established computer manufacturers look at the computer systems sold out of computer stores and sigh big sighs of relief. After all, computer stores

Approximate dollar price of computer systems

Price	Companies	Systems
10,000,000 – 250,000	IBM, Burroughs, NCR, Univac, CDC, etc.	"Mainframe" computer systems
250,000 – 20,000	Digital Equipment Corporation, Data General, Hewlett-Packard, etc.	"Minicomputer" systems
20,000 – 1000	Alpha Micro Systems, Apple Computer Corp., Vector Graphic, Processor Technology, etc.	"Microcomputer" systems

The theoretical three tiers into which the computer industry can be divided. In reality, many products fall into two of the classifications, while continuous improvements keep altering products' classifications over time.

sell primitive little systems to very small companies that couldn't afford anything better. But oh! How short-sighted these people are! Before 1990, microcomputer systems sold out of computer stores will account for ninety percent of the entire computer market.

The problem with computer professionals is that they are too quick to believe their own propaganda. In theory, the computer industry is divided into three tiers — mainframe, minicomputers, and microcomputers.

These three tiers of the computer industry are supposedly separate and distinct. You buy the smallest, least expensive mainframe when the largest, most expensive minicomputer will not do. And you buy the smallest, least expensive minicomputer when the largest, most expensive microcomputer will not do.

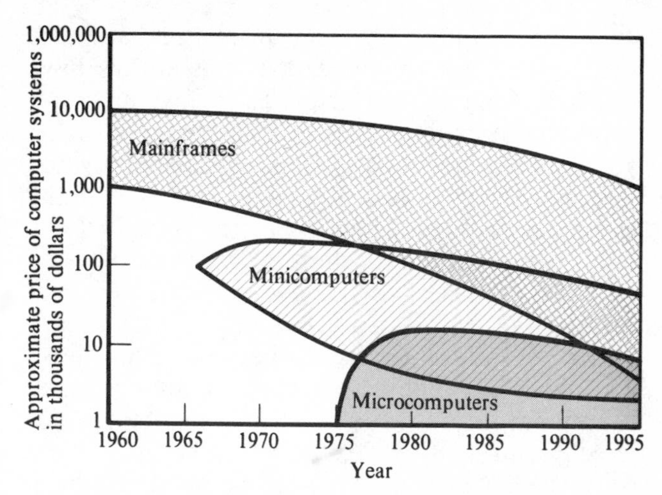

The computer industry divided into three tiers according to price. However, price can be misleading: it does not always reflect computing capability.

Most computer professionals agree that there is some overlap of products at the tier boundaries. But in truth, it is often hard to tell what type of computer system you are buying. In 1969 you would have paid $150,000 for a "minicomputer" system, equivalent to a 1979 $20,000 product, which may either be a "minicomputer" system or a "microcomputer" system. In 1979 you will get, for $150,000, a "minicomputer" system with all the capabilities and the appearance of a "mainframe."

"Mainframe," "minicomputer," and "microcomputer" are rapidly becoming labels. The product you buy is a "mainframe," a "minicomputer," or a "microcomputer" because that is the label bestowed on it by the manufacturer. We might therefore look at computer system price tags to see how the industry shapes up.

But computer system price tags are also misleading, because price does not always have much bearing on performance. When buying a computer system, it is very unwise to assume that you get twice as much computing capability when you pay twice as much money or that half the computer comes for half the price. Thus, in 1980, while there is still a substantial difference between the price of the least expensive "mainframe" computer systems and the most expensive "microcomputer" systems, they will likely have comparable capabilities. Electronics to make microcomputers as powerful as mainframes already exists. By 1990, there will be little or no difference, other than price, between products sold as "mainframes," "minicomputers," or "microcomputers." That is why, even though significant price differentials may remain, ninety percent of all computer systems sold will likely be "microcomputer" systems or their equivalents. And they will most likely be sold out of computer stores. The remaining ten percent will be the largest, most powerful, and most expensive of the "mainframe" systems; they will be needed to service a variety of specialized applications. Here are three examples of applications only a "mainframe" could handle:

1) Difficult computations associated with weather forecasting and geological data analysis;
2) Control of large information banks;
3) Data processing for state and federal government agencies (for example, the Social Security Administration).

But even the small residual market for large

computers will be dominated by new companies. Cray Research Inc. and Amdahl Corporation stand the best chance of dominating this small market.

On first inspection the computer revolution may look like one of the biggest job creators the computer industry has ever seen. That is partly true. Hundreds of new, small computer manufacturers have created an overwhelming market for electrical engineers and computer scientists. Much of this demand, particularly for electrical engineers, will remain and even increase over the next twenty-five years. But the future is not so rosy for computer programmers.

Wags like to point out that everyone in the world will have to become a computer programmer if we are to support the vast numbers of computers that everyone predicts will be sold. But within twenty-five years there may well be almost no programming jobs left. And the number of eliminated programming jobs will greatly exceed the number of newly created engineering jobs.

In order to understand why programming jobs will be eliminated, we need to look at how a computer is taught to perform its required tasks.

A computer's task is defined by the person who wants to use the computer. But that person usually has no idea how a computer works. For example, suppose you want a computer to process your company payroll. The company comptroller or accounting supervisor will define exactly how the payroll data is collected and how payroll records and paychecks are to be printed back. Then they throw their hands up in the air because the computer cannot read their definition of the problem, and they do not know how to talk to the computer.

Enter the systems analyst and the programmer. The systems analyst is a high "muckety muck" in the priesthood that surrounds computers, making them incomprehensible to all but the priesthood. The programmer is a junior functionary within the priestly ranks. The difference between these two is that the systems analyst understands how computer systems work, while the programmer merely understands a computer language. So the systems analyst takes your problem (which he usually understands partially) and lays it out in a form that makes best use of your computer's capabilities. The programmer then takes the systems analyst's design and converts it into an incomprehensible list of computer commands.

There are two problems in this sacred rite:

1) The program created by the computer programmer usually contains numerous errors no one except the computer programmer knows about because no one except the computer programmer ever looks at his program.
2) The program does what the systems analyst thought you wanted, and that may differ vastly from what you really wanted.

What ensues is an extended test of the limits of human relationships. For the next two years you wait in anguish while the computer does nothing and then does something that is incorrect. And the original three month estimate to get the entire job done drags on, and on, and on.

Here is the problem: the qualities that make

a perfect systems analyst or computer programmer are that he/she be absolutely accurate, remember the entire problem at all times, never sleep, work with lightning speed, and do exactly as he/she is told. If you know of people with these characteristics, a lot of potential employers would like to hear about them.

But these are the exact characteristics of the computer itself. Why not let the computer write its own program?

That is exactly what a number of companies are beginning to do. Computer manufacturers are designing "self-programming" computers, which you program by example, in English or the language you understand. No companies yet manufacture truly self-programming computers, but four who have taken small steps in the right direction are Logical Machine Corporation (also known as LOMAC), the Diablo subsidiary of Xerox Corporation, Qantel Corporation, and Microdata Systems.

Anyone trying to build a truly self-programming computer is faced with an appallingly complex task — one which computer manufacturers have had neither the time, the experience, nor enough inexpensive electronics to solve. But the problem can be solved, and for data processing computers it will be solved.

Self-programming scientific and engineering computers could be designed, but computer manufacturers may decide not to expend the necessary time and effort. Scientific and engineering computer applications are too varied, and scientists and engineers have gotten used to writing their own programs anyway.

The truly self-programming data processing

computer will let you define your problem by writing on the screen the information you have to input, and you will display on the screen the reports, checks, and other information you want returned. The computer will then ask you questions, via its screen, which you will answer at the keyboard, until all ambiguities in your problem definition have been completely resolved. The computer will then create its own program, to meet the requirements of your problem as explicitly defined by you. Such self-programming computers will likely be available within ten years. When they become available, everyone will be his or her own computer programmer.

The computer programming profession will then disintegrate into three small factions:

1) Systems analysts and programmers who design and build the self-programming computers;
2) People using computers for scientific and engineering applications;
3) People controlling operations in large companies.

We will discuss each of these three job categories in turn.

It takes a computer program — a very large and complex computer program — to turn an ordinary computer into a self-programming computer. Therefore, the companies that manufacture self-programming computers will hire systems analysts and programmers (but not very many of them) to design the self-programming characteristics of the computer.

Scientists and engineers will continue to program general purpose computers much as they do today. Engineers will also use computers as electronic components. For example, people who build robots will put computers into these products, but you could never tell it by looking at a robot. In products such as robots, the computer is an electronic "chip", which an engineer has tailored to his specific application by writing a suitable program. Most scientists and engineers write their own programs; they rarely use the services of programmers, so there is not much future for programming jobs here.

A large company cannot simply buy hundreds of small, self-programming computers, give one to anybody who needs it, let everyone do whatever he or she wants, and expect anything coherent to come out of the resulting mess. The moment any company has two or more computers on its premises, someone has to pay attention to the way in which these computers work together. This task is handled by systems analysts, whose jobs will therefore remain.

Creating a self-programming computer is conceptually similar, but far simpler than creating the "thinking" computer discussed in Chapter 3. In both cases, we interpose an intelligence between the human user and the elementary electronic logic of the computer. This allows the human user to define a problem by describing information entered and results expected, which certainly beats describing the problem explicitly as we did in the addition example of Chapter 3. But the self-programming computer is still a long way from a "thinking" computer. The self-programming computer might be programmed to give every appearance of intelligence, but it will still not be able to invent solutions. The self-programming computer will fail the moment you present it with information that it has not been explicitly programmed to handle.

So much for the impact computers will have on the jobs of computer professionals; consider next what computers will do in the average office.

You do not need to be an expert in microelectronics, nor yet a seer, in order to predict the changes computers will bring to the office. It makes little difference whether a company manufactures smog masks or sausages, sells advice or buys oil tankers; ultimately, company operations center on a bunch of people working in look-alike offices, with comparable office machines, doing fairly standard jobs.

Offices create information, store information and retrieve information. Also, they monitor money as it is received and spent.

Let us look at how these office operations will change in the coming twenty-five years.

Today most information is created by secretaries, using typewriters. Secretaries rarely use shorthand as a means of dictation anymore; they transcribe from tape recorders, which is far more efficient. A minor problem with typewriters is the fact that typists make mistakes. A major problem with typewriters is the fact that the typist's boss changes the text. Both of these problems are rapidly being solved by word processing systems, which started to bud on the market ten years ago and are now in full bloom.

Word processing systems store large amounts of text, usually on "floppy disks". A floppy disk looks like a 45 rpm record, permanently enclosed in a cardboard dust jacket. But the record is made out of thin plastic covered with the

Floppy disks look like phonograph records, but store information. They are soft and bend easily.

same type of magnetic surface that you find on a cassette tape. You can store as much as a million characters of information on one floppy disk. The word processing system displays text on a screen, allowing you to add, delete, insert, change, move, and variously format the text before printing it out on paper.

There are some very usable word processing systems on the market today, manufactured by companies such as Lexitron, Vydec, Lanier, Wang Laboratories, and Xerox. Of course, the design of word processing machines will continue to improve, making them even more usable in an office, but many are already very well designed and very usable. They are also quite expensive. A well designed word processing system, with a display screen, keyboard, printer, and a million character memory, will in 1979 cost anywhere from $8000 to $20,000. Within ten years equivalent systems will

IBM DF-2 Type is a unique
optical character recognition
type style combining
conventional characters
with a bar code

A B C D E F G H I J K L M N O P Q R S T U V W X Y Z
a b c d e f g h i j k l m n o p q r s t u v w x y z
> < # $ % Δ & * ( ) Λ ∘ - + ⌶ ? / ' ⑁ : ; , .
1 2 3 4 5 6 7 8 9 Ø

Electronic Reading Machines use an Optical Character Recognition (OCR) font.

be available for $300 to $1000 — the cost of a good electric typewriter. And a secretary with a good word processing system can do the work of two or three secretaries with standard electric typewriters.

A lot of information that gets typed comes from text that is already either typed or printed. Electronic reading machines are already available, but they are expensive and reliable or inexpensive and unreliable. Electronic reading machines sometimes require the text to be typed or printed using a special font, or character set, that makes the text more readable to the electronics but less readable to humans.

We will see dramatic advances in electronic reading machines. Within five years electronic reading machines costing less than $500 will be able to read anything printed or typed.

The ultimate typewriter will automatically type from dictation, since it will be able to hear electronically. As described in Chapter 4, Bell Laboratories has developed low cost electronics that hears, even allowing for accent variations. This electronics will be inexpensive and commercially available within five years, at which time it will be possible to combine a typewriter with dictation equipment. You will be able to dictate and have the typewriter automatically type back your dictation. Of course, the next advance will be to combine dictation equipment with a word processing system, which will let you fix the inevitable errors resulting from misheard words.

A word processing system that can automatically transcribe dictation will significantly reduce the number of typists' jobs in an average office. It will also pressure office professionals into using such equipment themselves — thereby eliminating typists entirely.

Reproduction technology is moving at a very rapid pace. Black and white reproduction was pioneered by the Xerox Corporation; today the copy machine is a standard piece of equipment in every office. Color copiers are available today, but they are expensive. They will not be expensive much longer.

Within the next five years photocomposing machines will be as inexpensive as any copy machine, and the office copier will be as efficient and economical as any printing press. We have already seen a proliferation of small offset printing offices such as the national chain Postal Instant Press. Offices, even small ones, will soon have their own in-house printing presses, since printing and office copying will merge into the same machine — which will also serve as a computer system's printer.

But when push comes to shove, microelectronics and automation will not have a dramatic impact on office jobs. Fewer secretaries will be needed — but they are in short supply anyway. Fewer file clerks will be needed, and low-level office positions may be eliminated. Office jobs will be more demanding, and office personnel will require more education, but there will be no significant decline in the number of jobs.

Possibly the most interesting aspect of the future office will be the way information is stored and retrieved. What makes this aspect of the future office so interesting is the fact that electronics will offer more capability than anyone could prudently use.

You can store anything in a computer's memory, from simple text to full-color pictures. Within five years an inexpensive office computer will be able to store a copy of a Rembrandt painting or an office letter, reproducing either on demand in black and white or in full color. Thus, the office computer's memory will hold everything and anything you can put on paper: letters, books, tables of numbers, blueprints, or art. Every office desk will have its own computer, with keyboard and display screen, and given an adequate index, you will be able to instantly seek out anything from the office files. You could as easily look up an individual piece of correspondence, or all correspondence and blueprints covering a specific design project, or all pictures painted by artists between 1940 and 1955 with bluebirds in them.

The information that will be at your disposal in the future goes far beyond that which can be stored in your own local office computer. In fact, one of the industries with great growth potential during the coming industrial revolution will be the information industry. There are already many companies the principal business of which is maintaining computer data bases. (A data base has a large quantity of information, which is stored in a computer-readable form, with appropriate indexes that a computer can use to find material that you request and identify by subject matter.) For example, an attorney's data base holds extensive legal records. Computers can search through this vast data base, selecting cases that may be relevant to any action that an attorney is currently working on. Doctors, lawyers, teachers, and professionals of all kinds will form an enthusiastic and lucrative customer base for companies, present and future, that maintain data bases. And conversely, the office worker of the future will be able to obtain information on almost any subject instantly via an office computer terminal.

The office worker will be able to generate reports at will. A report may be a simple compilation of data from the computer's data base, or it may be a complex financial analysis. Within twenty-five years computer programmers will have been eliminated; you will be your own computer programmer. With no programming priesthood separating you from the computer's data base, your ability to retrieve information and have reports generated electronically will be almost unlimited.

Let us examine the effect that computer maintained data bases could have on the legal, medical, and teaching professions.

The microelectronic revolution is already changing the legal profession. Computer services such as Westlaw have given lawyers low-cost access to legal information, which in the past would have been prohibitively expensive to compile. This well established trend will grow to the point where computers provide attorneys with nearly all their case preparation. This might jeopardize many legal assistants' jobs, but it can only improve the lot of the attorney. Humans still have to take the information computers have compiled and manipulate this information to best support each side of every dispute. Computers can never replace humans arguing cases or making judgements for the same reasons that computers will not, in the foreseeable future, show human intelligence (as discussed in Chapter 3). The law never has been so definitive that a computer could decide any case on its merits.

The fact that a dispute reaches the courts means the interpretation of law is being questioned. Resolutions are invariably based on human judgement, something computers do not have. If legal disputes could be settled definitively, based on absolute right and wrong, computers could take over from judges. But under such circumstances there would be no litigation in the first place.

Widespread use of computers within the legal profession may well cause law firms to start linking together in national confederations. Given electronic mail, the telephone, and computer data bases, it will be equally easy to deal with a co-worker who is in an adjacent office or in an office on the other side of the country. Thus, microelectronics will have a greater impact on the structure of legal firms, and the way lawyers work, than it will on the position of lawyers in our society.

The medical profession will be impacted more severely than the legal profession by the microelectronic revolution. Doctors spend much of their time discovering a patient's symptoms, looking at the results of diagnostic tests, then trying to solve an intellectual puzzle: what is the most probable cause for the symptoms? And having selected the most probable cause, what is the most likely remedy?

A computer might be more accurate than a human when it comes to selecting possible causes that match the symptoms and selecting proper medication for the assumed cause. This is because the computer could search an entire data base, covering many thousands of common or obscure possibilities, omitting nothing, and including the most recent findings in its search. Likewise, the computer would not overlook any possible remedy when selecting the most desirable medication.

Doctors already use computers to aid in diagnosis and to help select medication. There will be a big increase in such use of computers. For each patient, the doctor will input symptoms and medical test results and receive from the computer a list of most probable diagnoses and courses of treatment. The doctor will then use his/her judgement to select a diagnosis and course of treatment.

But computers may potentially cause as many problems for the medical profession as they solve.

The medical profession, like every other profession, includes people who fail to maintain their training. But in the medical profession these people can be particularly dangerous because of their ability to escape detection. In most other professions, people work as part of a team, in which case peer evaluation identifies incompetents and isolates them into positions where they will at least do no harm. But the incompetent doctor can withdraw into a private practice where he/she sees only patients who do not have the expertise to judge the doctor's qualifications. Will medical data bases, as they become more efficient, accelerate the deterioration of doctors' competence, much as electronic calculators are eroding our ability to do mental arithmetic? This is a problem that the medical profession should take very seriously; it has many ethical and legal implications.

Then there is the problem of computer misuse. A computer will serve any master with equal dedication. As computers become more capable, they also become more potent as tools for anyone wishing to use or misuse the capability. We would all applaud the ability of a medical data base to provide the competent doctor with mind-jogging tidbits that lead to early diagnosis of dangerous ailments. But we should be equally afraid of the ease with which untrained personnel could access the same information. What is to stop the widespread use of marginally qualified medical technicians using computer data bases to perform medical diagnoses for every employer, insurance company, or miscellaneous authority that presumes to make use of such information? And what about the profusion of charlatans and questionable characters who lurk on the fringes of the medical profession, selling services that are ineffective, questionable, or downright illegal? A medical data base would serve them as willingly as it would serve the most competent doctor.

Every use of computers implies a possible

misuse. But we must worry more about misuse of computers for medical purposes because so many medical quacks don the mantle of messiahs, while illness and healing are human concerns too highly charged to tamper with lightly.

Computers and microelectronics have already revolutionized hospitals (and, some would claim, have contributed to the high cost of hospital care). Microelectronics has found its way into almost every piece of medical equipment, from costly three-dimensional tissue scanners down to thermometers and the most inexpensive instruments. Computers are used to monitor patients and provide nurses with information as they make their rounds. We do not have to tell doctors and nurses about the microelectronics revolution: they, more than anyone, can tell us about it.

The impact of the microelectronics revolution on education and educators is very enigmatic. The microelectronics revolution will place a tremendous new responsibility on the shoulders of educators, who will have to cope with job retraining and lifelong learning. Will microelectronics and computers help solve the problem they create?

Computer aided education has been around for a long time, yet I doubt if there is any place where computers have offered so much potential and delivered so little product. In fact, "Sesame Street," the pre-schooler's television program, probably represents the most effective use of electronics in education that we have seen up till now.

PLATO is an educational data base pioneered by Control Data Corporation. It is a thing of beauty and awesome capability, yet its impact on education in general has, until now, been insignificant. This is probably because of the high cost of both the special computer terminal needed to access PLATO and the subsequent computer time. But this high cost barrier will surely tumble before the onslaught of the microelectronics revolution. If it were possible to access PLATO via a home television set, or a computer terminal of comparable low cost, and if subsequent computer time charges were nominal, it is hard to predict how profound the impact of PLATO on education might be.

We can expect to see important changes during the next twenty-five years in the educational system, and in the tools available for improving that system.

Consider first the education problem that must be solved. The world has, and will continue to have, a very uneven distribution of qualified teachers. In the developed and industrialized nations they are in oversupply. Yet there is a desperate undersupply of qualified teachers in Third World nations. There is also a massive difference in the job educators must accomplish. In the least developed nations of the world, educators must still strive to achieve the elusive goal of providing everyone with a basic education. At the other extreme, industrialized nations will be faced with a whole new problem: retraining workers whose jobs are eliminated by automation and providing other workers with continuing education so they can cope with changing jobs.

In underdeveloped parts of the world, where there simply are no teachers, one would expect that any form of electronic education would have to be better than nothing. Yet this has not been the case. Although radio has been with us for more than sixty years, and television for perhaps forty, neither radio nor television play the significant educational role in underdeveloped countries that many predicted. Trained teachers use radio and television successfully as aids, not substitutes. But the microelectronics revolution will, for the first time, allow television to actually substitute for a teacher when no teacher is available. Why? Because students will, for the first time, be given the choice of learning at their own pace.

A program broadcast via radio or television has to be watched, in its entirety, at the time of the broadcast. If the broadcast spends too much time on a subject or skips over material too quickly, the student has no recourse. Given one broadcast and many students, it is a truism that most students will find different parts of the broadcast too lengthy or too slim. But worst of all, students subsequently have nothing to rely on but their memories. In the future, very long educational television programs will be broadcast to remote corners of the world, to be recorded locally on video cartridges or video disks. Video cartridges and disks are already widely sold as attachments to domestic television sets. They allow many hours of television programming to be recorded and selectively played back through a television set. Video cartridges and video disks allow students to choose the material they wish to view, skipping subjects they understand, while repeating as often as they wish material they have not yet learned. Moreover, the recording is permanent. Students can go back to it a day, a week, or a year later.

The microelectronics revolution will allow students in remote corners of the world to have two-way communication with teachers who might be half a world away. Via microwave communications it will be possible for teachers and students to converse. Thus, students could ask questions and receive immediate responses.

Control Data Corporation's PLATO system, if it could be distributed around the world at very low cost, might do more to educate the people of underdeveloped nations than any other program or plan.

With the exception of any impact resulting from a low-cost PLATO system, the microelectronics revolution is unlikely to have a significant impact on general education in industrialized countries. High school curricula will change to include more technological subjects, and computers will be used in the classroom as teaching aides, but microelectronics promises no radical changes in teaching methods. Already computers are being used in language drills, mathematics, and of course, computer programming courses. We can expect to see computers used more as encyclopedias and as means to drill students in a wide variety of subjects.

Computers will help broaden the range of subjects that can be taught. Despite the surplus of qualified teachers in most industrialized countries, isolated shortages will exist. Thus, at Stanford University obscure languages are taught entirely via computer programs. A student can learn a west Armenian dialect, even if no member of the faculty understands a word of the language.

A different problem will cause teacher deficiencies in high technology areas themselves. Teacher salaries do not vary a great deal with the subject matter being taught. In high technology areas this causes problems, since anyone who knows enough microelectronics to teach the subject can earn two or three times a teacher's salary as an engineer or scientist. In consequence, few qualified teachers of electronics are likely to stick around the teaching profession. Here again, the only solution will be PLATO, or well-planned courses that rely on television and video cartridges as a substitute for the teachers who changed jobs once they learned enough technology to be engineers.

And what about lifelong learning? Clearly we will need more teachers (particularly at the university level) in a society where professionals must go to school indefinitely. The challenge will be to keep our teachers ahead of their students (and to keep them teaching). Already in 1979 more faculty in electrical engineering and computer science departments than we would care to admit are no longer qualified to teach the courses to which they have been assigned. And as the microelectronics revolution spreads into disciplines far removed from microelectronics, the problem of teacher obsolescence will follow close behind. This will further aggravate the shortage of adequately trained teachers. And at that point we will be no better off than students in underdeveloped countries. Amusing as it sounds, we will have to apply the same techniques to educate illiterate children in underdeveloped countries, where there are no teachers, as we will to educate highly trained professionals in developed countries, where there are no adequately trained teachers.

Let us now examine the impact that microelectronics may have on a broad range of

Photo courtesy of Radio Shack, a division of Tandy Corporation.

A home computer. In the future, terminal networks will extend beyond the office into the home, and this, more than any other development, will eliminate many white collar jobs.

white collar jobs, as compared to the three specific professions we have described.

Computer terminal networks will extend beyond the office into the home. This development, more than any other, will eliminate many white collar jobs. Let us examine the reasons for this.

Any television set can serve as half of a computer terminal — the display half. The other half consists of a keyboard, much like a typewriter keyboard, plus assorted electronics. In 1979, this keyboard and electronics may cost anywhere from a couple of hundred dollars to perhaps a thousand dollars, depending on how fancy you want to get.

If you have an electronic game hooked to your television set, then you are probably already using it as a computer terminal, since most electronic games have a microprocessor inside them. But what makes your television set a really powerful computer terminal is the fact that you can connect it to anyone else's computer — with a telephone call. Just as you dial a telephone number to talk to a friend, so you can dial a computer's number and have your terminal converse with the computer. The electronics that connects your television set to a telephone is inexpensive and available today. The problem is that your telephone bill would be unacceptably high were you to use your television set extensively as a terminal, communicating with a variety of computers over telephone lines. But telephone and microwave communications probably represent the biggest growth industries of the next twenty-five years, and as use increases, costs will decrease. We could easily see a tenfold increase in the size of the communications industry, and this growth will be accompanied by massive cost reductions. Then it will be economical for the television set in every house and the computer terminal in every office to spend one or more hours of every day communicating with a variety of computers via telephone calls.

And this has some very unexpected implications.

Consider how many white collar jobs involve nothing more than providing information over the telephone. A stockbroker's primary function, for example, is to execute customer buy and sell orders. The stockbroker's secondary function is to give customers current stock prices and performance histories. Stockbrokers also give customers advice as to stocks they should buy or sell. A University of Chicago business school study showed that only in rare instances does advice from a stockbroker beat random selection.* Most stockbrokerage houses offer full services, with research departments and buy/sell advice, but a growing number of small stockbrokerage houses now offer buy/sell services only, with no advice, no research departments, and lower brokerage fees. Today active stock traders can, and do, buy their own computer terminals, which they use to obtain stock price quotes on demand. Such terminals cost less than $1000. Within the next ten years television sets will function as terminals in your office or home anyway; using the terminal to get information on stocks, to buy or sell, will then become an almost free extra service. You will be able to access a stock market data base to get information you want, be it price or past performance of any stock. You will even be able to receive your stockbrokerage's recommendations via your home or office computer terminal. You will also be able to place your own buy and sell orders. There is nothing left for a stockbroker to do that a computer terminal could not do. The live stockbroker does not even offer additional security against fraud. Stockbrokers take the bulk of buy and sell orders over the telephone, recognizing the customer via his or her telephone voice, and executing the order without written confirmation. The computer might instead demand the customer's written signature which, transmitted as a picture, could be compared with a stored sample. And if that is not good enough, a customer's fingerprint could be transmitted and compared. (Law enforcement agencies today use computers

* See Burton G. Malkiel's *A Random Walk Down Wall Street*, Norton Press, 1973.

to check sample fingerprints against fingerprint data banks for known felons.)

So stockbrokers will go the way of the blacksmith, to fond remembrance.

Of course, some personnel must remain in stockbrokerage houses to oversee operations and check out special circumstances. Supervisory and management positions will remain, but run-of-the-mill stockbrokers will no longer be needed.

The U.S. Securities and Exchange Commission is experimenting with an entirely electronic stock market. At the small Cincinnati Stock Exchange, all buy and sell orders are matched by computers, which also adjust stock prices up or down in order to keep buy and sell orders balanced.

Will all stock exchanges become entirely electronic operations? They could, but such a move would be a tragic mistake because the risk resulting from computer abuse is greater than the rewards offered by correct use. We discuss this problem in Chapter 7.

Computer terminals in every office and home will eliminate any job that simply entails responding to information requests or executing routine orders. Here are a few segments of the work force that fall into this category:

1) Telephone directory inquiry personnel;
2) Airline information and reservation personnel (this will apply also to bus and train service personnel);
3) Central telephone operators in government or industry offices (frequently known as PBX operators).

Office and home computer terminals could

easily handle any of the inquiries described above. You will connect your terminal to the appropriate computer by dialing the appropriate telephone number (or you may type in the number at your keyboard). Questions will be displayed on your screen, and you will respond as directed in order to obtain information or place orders.

Handicapped individuals would have little trouble with computer terminals. The blind, for example, might use a braille keyboard and have a voice generator to replace the television screen. Quadriplegics could rely on electronic hearing devices to enter information into the computer.

There are a variety of service professions that could be eliminated, along with inquiry and order placement jobs in general. For example, consider travel and employment agencies.

The primary function of a travel agent is to know how to decipher the blizzard of routes and fares connecting any two cities in the world. They may also issue tickets. But a computer is far more efficient at selecting routes and finding the lowest fares. In fact, the computer is used by travel agents today to identify the viable ways of flying between two cities in the world and to display fare options. By connecting this service to home computer terminals, we could eliminate travel agents.

There are, in fact, a number of advantages to handling travel reservations via your own computer terminal and a central data base. Airlines are plagued by "no shows", the businessman who honestly changes his plans and forgets to tell anyone, or the dozens of bogus reservations some airline personnel create in order to guarantee standby space for themselves. Both cause airlines to fly partly empty airplanes or to overbook —

which can get airlines into worse trouble on those rare occasions when everyone does show up. But the computer could keep track of "no shows", tracking them to the terminal at which the reservation was made. This would make it possible to demand non-refundable deposits or enforce other penalties for uncancelled reservations.

Travel agencies also prepare package tours. A package tour is nothing more than a sequence of block reservations with airlines, hotels, resorts, and ground transport. By obtaining lower prices through block bookings, the package can offer an attractively priced excursion. Here again, the computer can do a very efficient job of piecing tours together. Any hotel, resort, airline, or ground transportation agency that wished to participate in packaged tours could enter into the computer data base their availability and cost. The computer could then piece together tours, doing the job far more effectively and with greater versatility than any human. You could, for example, tell the computer where you wished to go and for how long. The computer would piece together various low-cost options, allowing you to select which, if any, suited you.

A computer could even provide descriptions of places and helpful travel hints. Anything a travel agent may tell you over the telephone, the agent could as easily enter into the computer as a permanent record. For that matter, travellers who used the computer could add their own comments, allowing future travellers to review the experiences of those who have gone before.

Employment agencies, likewise, could be eliminated by a jobs data base. People looking for work could enter their resumes into the data base via their home computer terminal, together with a photograph of themselves or a video recording in which they discuss their qualifications and objectives. A potential employer might describe the job and candidate qualifications that he/she is interested in, then obtain a list of names from which individuals could be selected for more detailed review or personal interview. Employment agencies offer no more, and frequently a great deal less, than this.

If yours is a service-oriented job that involves giving people information and/or placing their orders, then you should carefully evaluate how much professional expertise you really do give your customers, and how much of your professional expertise is really make-believe. Because if it is really make-believe, your job will be gone within twenty-five years.

Many of the predictions in this chapter assume that every home and office of the future will have one or more computer terminals.

We can argue about the way in which computer terminals will be used in homes and offices, but there is no argument that homes and offices will all have computer terminals. It is already happening — particularly in Europe. The trend in Europe began in Britain, with a system called Viewdata, which transmits written material via telephone lines to television sets all over Britain. Any Briton whose television set is appropriately equipped can read news bulletins or the weather forecast; he or she can buy a variety of products or use various services. In short, he or she can already do most of the things described in the preceding pages.

6

Industry, Evolution, and Revolution

We have described how microelectronics will affect jobs and professions over the next twenty-five years. A more difficult task is to examine the impact that microelectronics may have on specific industries. This impact includes obvious evolution, as seen in the design of automobiles, obvious revolution, as per electronic calculators, and the birth of brand new industries, such as replacement of human parts and the day of the bionic man.

Consider the automobile industry. Automobile designers are beginning to use microelectronics in a variety of ways. Three examples are driver instrumentation, fuel mixture control, and anti-skid braking. In consequence, the consumer gets an automobile that is easier to drive and has better fuel economy, but is otherwise simply next year's model.

Even though microelectronics will not

revolutionize the automobile per se, it will revolutionize the way in which automobiles are manufactured, by allowing robots to replace people on the production line.

Other industries will experience the type of revolution suffered by the mechanical calculator and watch industries. And in all probability, the pattern of revolution will repeat itself: existing manufacturers will not anticipate the revolution and will be upended by microelectronic products that come out of the semiconductor or related industries — a quarter to which they never previously looked for potential competitors.

Unpredictability is one of the key ingredients of revolution. Ergo, it is hard to predict where the revolution is likely to manifest itself next. Of the many possibilities, we will therefore confine ourselves to two probabilities: the recorded music and the camera/film industries. We will also look at the impact microelectronics may have on publishing and at the possibilities for bionics.

Consider first the recorded music industry. Studios contain a lot of expensive equipment that records sound on records or tape cassettes. To play back records and tape cassettes, you need expensive, high-fidelity record players or tape decks.

There are two serious deficiencies in this scheme of recording sound and playing it back: the recording method itself is not very good, and the problems are aggravated by moving parts in the playback system.

Studios record sounds as signals on magnetic tape, or as wiggles within the groove of a record. Both are poor recording media; they generate background noise and are easily damaged. Much of the expensive electronics in your stereo system is devoted to suppressing this unwanted noise while preserving the wanted sound.

Additional expensive electronics is required in order to keep the tape or record moving at some very precise speed. Moreover, the actual movement of tape against read head or of needle within groove generates fidelity problems that are aggravated by dirt and adverse environmental conditions.

Microelectronics can resolve these problems by storing sound as a sequence of numbers.

Sometime between the years 1990 and 1995 records and magnetic tape will start being replaced by microelectronic memory chips. Sound will be stored on these memory chips by "digitizing" the wave form that would otherwise be held on record or tape cassette. The continuous signal used by the recording industry today is called an analog signal. The digitized recording technique of tomorrow is called digital recording.

Digital sound is recorded as a sequence of numbers, each of which can be visualized as representing the distance of a step from a zero base line.

Digital recording offers a potential for higher sound fidelity than analog recording. This is because numbers are either right or wrong, whereas an analog signal is never read back absolutely right, and therefore it is always slightly wrong. In fact, the quality of your music system is largely based on the ability of the system's electronics to minimize errors in the analog signal it reads back.

The fact that numbers are either right or wrong is great — provided the numbers are right. Then read back electronics can recreate a completely error-free signal. But what if the numbers

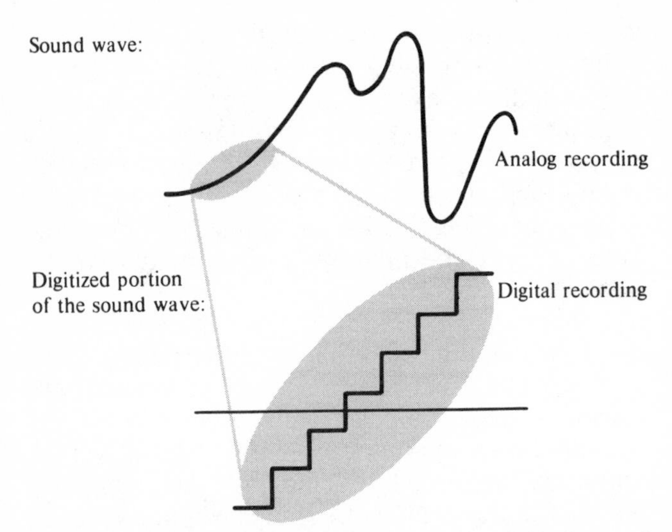

Sound waves can be recorded digitally. When the individual steps of the digital recording are short enough, the sound produced cannot be distinguished from the actual sound wave.

are wrong? Fortunately, numbers can be recorded with companion error correction codes. These codes allow read-back electronics to identify numbers that are wrong; in most cases read-back electronics can correct the errors.

A number of companies are experimenting with digitally recorded sound today. They record sound on records and magnetic tape as a sequence of numbers. In Britain, the government broadcasting networks have been experimenting with digitally broadcast music.

Recording sound digitally on a record or magnetic tape does not solve fidelity problems resulting from record or tape movement. And many audio engineers claim that they will be able to improve analog recording and reproduction techniques fast enough to stay ahead of digital

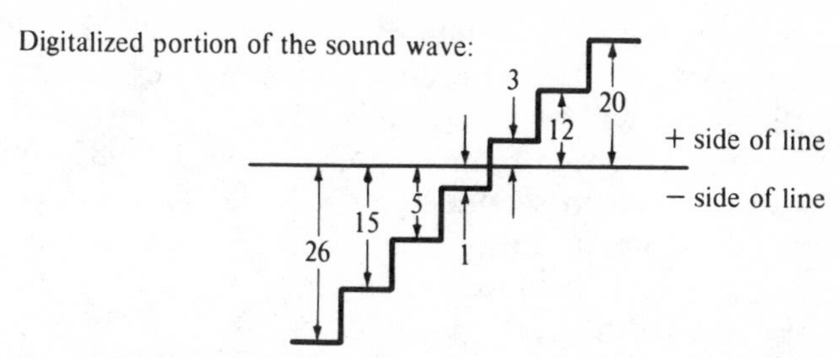

Recorded digitally as follows:

-26, -15, -5, -1, 3, 12, 20

Digital sound is recorded as a sequence of numbers. Each number can be visualized as representing the distance of a step from a zero base line.

recording. But microelectronics will ultimately replace records, magnetic tape, and moving parts in general.

Between the years 1990 and 1995, it will be possible to record half an hour's worth of music (or more) on a semiconductor memory chip that is smaller than your fingernail. This chip, mounted in some suitable holder, will plug right into a socket in your new high-fidelity music system, silently pouring out its stream of numbers into the circuitry that drives speakers. Not a single moving part will remain.

We can record digital music on memory chips today. It takes approximately 65,000 binary digits (otherwise known as bits) of memory to record one second's worth of music. That much memory (of the correct variety) is available today for two or three dollars. Thus, we could economically sell one second's worth of sound on a memory chip for the retail price of today's average long playing record or tape cassette. And that is not going to cause a stampede to record stores. But every two years, the semiconductor industry quadruples the amount of memory available for the same price. Therefore, somewhere around 1981, you will be able to buy four seconds' worth of sound for two or three dollars, and by 1983 you will get sixteen seconds, and by 1985 you will get roughly one minute, and by 1987, you will get roughly four minutes. You can take the calculation on from there.

Some experts argue that such indefinite extensions of memory chip capacity cannot be made. In particular, some experts claim that we will soon reach limits that prevent any further economic expansion of memory chip capacity. (Arguments against the existence of any such limits are given in Appendix A.)

When it is economical to record and reproduce music digitally using microelectronic memory chips, the average sound enthusiast will be able to buy recording equipment that is equal to the best found in any professional studio. The quality and fidelity of sound reproduction will be limited only by speaker technology — for which microelectronics can do little or nothing.

Although digital sound is the technology of the future, electronic music has been around for many years. In fact, Moog synthesizers, electric organs, and electric guitars were developed in the fifties and sixties. Today music synthesis represents one of the more popular pastimes for microcomputer enthusiasts. Microcomputer enthusiasts create sound mathematically, and a listener does not need much of a musical ear to differentiate between an orchestra and computer synthesized sound. I do not see a big future for computer synthesized sound. It is an interesting curiosity with a limited following (particularly if you like listening to cats yowling). But digitally recorded music does not synthesize sound waves — it records sound exactly, more faithfully than the best record or tape cassette.

Microelectronics and digitally recorded music will make possible two extensions of the recorded sound business that are not immediately obvious. These include voice enhancement and automated composing.

It is already standard practice in recording studios to electronically enhance the voices of singers whose names are better known than their voices would justify. Electronics that enhances the human voice will likely become a standard feature of all home recording equipment. This will have a profound effect on the recording industry since there will be little discernable difference between the recording made by a truly talented singer with a superb voice and the recording made by an ungifted amateur who can do little more than hold a tune.

There are two ways in which voice (or for that matter any other sound) can be enhanced. By analyzing the sound wave patterns, it is possible to identify components that contribute to pleasing sound and components that do not. A computer could doctor the incoming sound wave and enhance it prior to recording. A more complete enhancement might come via substitution. The sound waves created by someone singing or playing a note on an instrument might act as an index to equivalent, high quality sound waves held in a computer's memory. The high quality sound waves taken from memory would be recorded. In other words, your voice might be continually replaced by that of a well-known singer, or a combination of well-known singers, for whom complete libraries of voice sounds might be maintained in computer memory.

Composing music with the help of a computer presents a more interesting possibility. You will be able to compose music at your home computer and have your score instantly played back through your hi-fi system. Every note of every musical instrument that one might encounter in an orchestra or band will be digitally recorded for instant recall. You will create scores at your computer display. You will assign instruments to each portion of the score, then with a touch of a switch have your hi-fi system play back an orchestral or band rendition of your composition. Consider the fun you could have, once you have the score right, switching violins for electric guitars or oboes for saxophones. Such a machine could be built today, but it would cost more than a quarter of a million dollars, which would likely limit its sales potential. But within twenty years this equipment will be generally available, selling for less than $1000, as an add-on to home music systems.

The first computer composing systems will likely be crude. Every violin does not produce the same music, nor does every violin player. Moreover, the same note will produce a different recording depending on where it occurs in a musical phrase. We might guess that the first compositions synthesized by computers from pre-recorded sound libraries will sound lifeless. But in the long run, the computer offers far greater versatility than has ever been available from instruments in the hands of humans. The computer will provide not only the ability to enhance sounds, as described earlier, but in addition it will allow you to

Photo courtesy of Honeywell, Inc.

Honeywell's Visitronic contains a circuit that, by comparing two images "seen" electronically, automatically controls camera focus. The photo shows the circuit, the module it's encased in, and one thin dime.

specify tone, note duration, or any other definable variation of sound.

So much for the music industry. Now consider what microelectronics will do to film and camera makers.

Cameras already contain a significant amount of electronics to control lens aperture as a function of film speed and available light. Honeywell has a circuit it calls the Visitronic, which automatically controls camera focus by comparing two images using crude "seeing" electronics.

The Visitronic chip is, in fact, less than a half centimeter long. This chip is available today and allows cameras to automatically focus themselves in any light, ranging from candlelight to a brightly

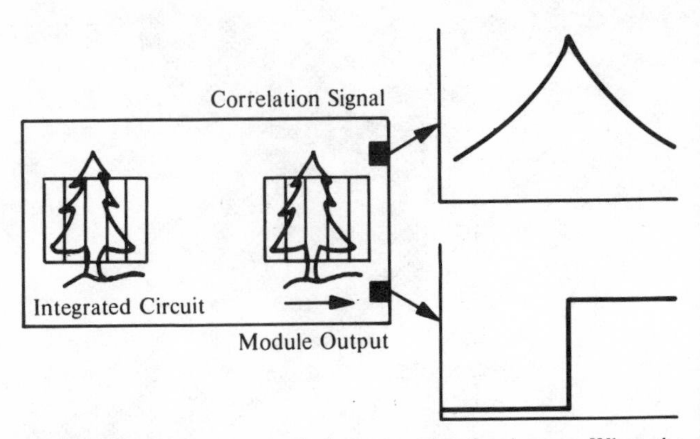

The auto/focus detector works by comparing two images. When the integrated circuit "sees" the two images as most similar, the correlation signal peaks. At that point, the module produces a step function voltage change that automatically positions the camera lens.

lit snow scene.

Today all cameras use film and the manufacture of film is a black art dominated by a few corporate giants. Microelectronics will eliminate film in twenty or twenty-five years.

Using microelectronics that is a logical extension of the Visitronic seeing panels, cameras will record pictures on a semiconductor "retina" capable of detecting color and light intensity, much as the human eye does. Electronics will convert the color and intensity detected at each retina point into a number that identifies the color and intensity. The entire picture is recorded as a large sequence of numbers. Subsequently, you print back a picture by feeding this number sequence into an appropriate printing device. The printing device contains electronics that converts the number sequence back into appropriate points of color and intensity, thus re-creating the picture. This is the same technique NASA uses today to transmit photographs from satellites to earth. The picture taken by a satellite's camera is converted into a large number of dots, each of which has color and intensity. The color and intensity of each dot is converted into a number, which is transmitted to earth. A printer re-creates the picture by interpreting the color and intensity from the received numbers.

A camera's images will be stored in removable microelectronic memory clips. A camera's microelectronic memory clips will be reusable. When a memory clip is filled with recorded photographs, you will take it out of the camera and insert an empty clip. Then you will take the recorded memory clips home and insert them into the color printer you use to receive mail. Within ten years, full-color printers of this type will be available for less than $1000, offering superb full-color fidelity. You will be able to print out as many copies of a picture as you wish and then save the camera's electronic memory clip to print more pictures in the future. Or you can reuse the memory clip. When you consider that color prints today cost at least twenty-five cents each, and a roll of film costs anywhere from three to six dollars, in approximately fifteen years electronic memory will be cheaper than film. Since a camera's electronic memory clips will be reusable, they could cost ten or twenty times as much as a roll of film and still be priced competitively. As for the home color printer, it will come almost for free since you will have to receive your mail anyway.

So farewell to the film and film processing industries.

The impact of microelectronics on the publishing industry is harder to predict. So to simplify the problem we will divide published material into that which must be portable and that which need not be portable.

If you buy something to read while traveling, it must be portable. Were electronics somehow to replace all published material, the electronic replacement would have to be equally portable.

But newspapers, magazines, and books, when read at home, in the office, or school, need not be portable, and an electronic substitute, likewise, could be based on the television set or a computer terminal.

When examining what microelectronics may do to, or for, publishing, we must also distinguish between competition and substitution. For example, one might argue that television competed directly with pictorial magazines such as *Life* and *Look* and put them out of business. But news-oriented magazines such as *Time* and *Newsweek* survived because their depth of coverage far exceeds anything provided by television. For the same reason, newspapers survive. Therefore we may conclude that television did not compete with newspapers and news magazines but rather augmented them.

Electronics is capable of directly replacing non-portable published material — which is not to say that it will. The replacement medium for non-portable published material will be the video disk, which is the next evolution of the video cartridges that are already widely sold as attachments to television sets. The video disk, rather than the video cartridge, will replace non-portable published material because the video disk lets you randomly select your viewing sequence, while the video cartridge must be viewed serially.

To understand the significance of the video disk, imagine an electronic newspaper that might be transmitted to your home overnight via a television broadcast channel, to be recorded either on a video cartridge or a video disk. If the newspaper is recorded on a video cartridge, then you will rewind the cartridge and begin viewing it at the beginning. There may, perhaps, be an initial table of contents that allows you to select the topics that most interest you. The cartridge will wind forward until it reaches the appropriate piece of tape upon which the topic you select is recorded. You will see pictures on your television screen with sound and appropriate text below the pictures. If you want to skip something, or you change your mind about the material you wish to see, the cartridge must spin backward and forward, locating the piece of magnetic tape that contains your next choice of information. This process is slow, and it leads to a lot of broken tape. The video disk is a far more useful recording surface because any point on the surface can be accessed with equal ease. The disk spins, and a reading mechanism at the end of a moving arm (called a read head) can almost instantly access any point on the surface of the disk. Now if you want to pick your way through the news stories on the video disk, you can do so with almost instant response.

There are two types of video disks: one is magnetic, like cassette tape, the other uses a laser beam.

Magnetic video disks, like magnetic tape, store information as magnetized zones. Magnetic video disks organize the magnetized zones in concentric tracks, whereas tape has parallel tracks running the length of the tape.

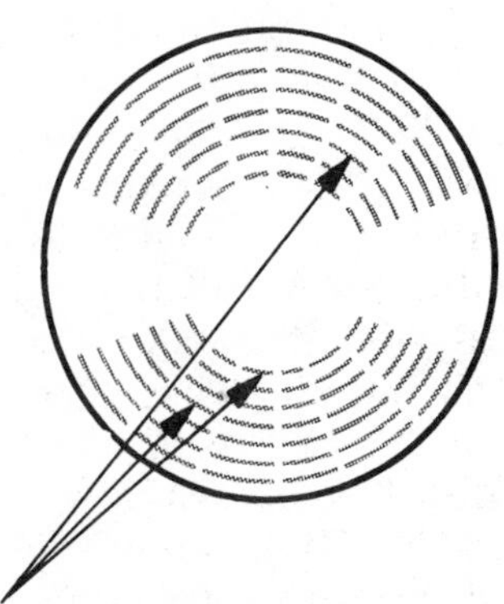

The video disk. The rotating disk is more convenient than a video tape cartridge because any point on the disk surface is quickly accessible to the reading mechanism, allowing you to randomly select your viewing sequence.

Laser disks are usually made of mylar, or some similar tough plastic, coated with a thin layer of aluminum. Information is stored on these disks by blasting microscopic holes in the aluminum coating using a laser beam. Subsequently the information is read back by bouncing a low intensity laser beam off the disk surface; the reflected beam is strong where there is no hole and weak where there is a hole.

The advantage of magnetic disks is that any home system can record information as easily as it can read the information back. Laser disks, on the other hand, can be read with a low intensity laser but must be written with a much higher intensity laser. Therefore, laser recorders will cost a lot more than laser readers. Also, you cannot erase or reuse a laser disk as you can a magnetic disk, because the laser disk has holes permanently burned onto its aluminum surface. Therefore you will acquire a laser disk much as you would a book, magazine, or newspaper — to use then store or throw away.

At present laser video disks are commercially available; magnetic video disks are not.

Television sets with video disk attachments may be capable of replacing non-portable published material, but portable published material will not be nearly as easy to displace electronically. A flat television set and a lightweight video disk system will have the size and weight of a small briefcase — which you will have to carry around if you want a portable system. In twenty or twenty-five years it may be possible to generate a package as small as a book, containing a display screen on the front and sufficient microelectronic memory in the package to hold an entire book, magazine, or newspaper, with illustrations. But this book size product will likely cost a hundred dollars or more. To be economically viable, its electronic memory will have to be reusable. You might, perhaps, plug it into a charging point at the airport book shop and have a newspaper, magazine, or book loaded into its memory. The charge would cost something, perhaps as much as the printed newspaper, magazine, or book would have cost. So why not buy the book?

The Xerox Corporation has a research team working on electronic books at the Xerox Palo Alto Research Center (PARC) in California. It calls its product the Dynabook. Dynabook is a very expensive experimental product today. It has a long way to go before it poses any threat to printed books.

Whether people will even bother with electronic books is far from clear. It seems almost as straightforward to buy a newspaper, magazine,

Photo courtesy of PARC Xerox.

The Xerox Dynabook — tomorrow's portable electronic book? The present Dynabook system.

Photo courtesy of PARC Xerox.

Mock-up of future portable Dynabook

or book in the first place. An electronic book such as the Dynabook will likely appear, but it will have its own specialized market; it will probably not displace portable published material in general. We might conclude, therefore, that microelectronics will augment portable published material rather than compete with it. Let us examine this in more detail.

Microelectronics may well replace printed reference materials and subscription periodicals. When every home has a television set and companion video disk recorder, cable television or other data transmission services will be able to give you low-cost access to a wide variety of newspaper-, magazine-, and book-type information. The advantages of an electronic substitute will be large enough to threaten a substantial part of the market for books, newspapers, and magazines in the home, office, and school.

If you subscribe to a newspaper, you will no longer have a roll of paper flung on your doorstep once a day. Instead, the newspaper will be transmitted during the night via a television channel, or cable television lines, to be recorded on a video disk. In the morning you will view the newspaper using your television set, picking your way among the pictures and stories recorded on your video disk, much as you would pick your way through a newspaper. But what will happen to advertising? After all, advertising is the principal source of income for newspapers. It is hard to avoid seeing an advertisement printed on a newspaper page. It would be easy to bypass advertisements on your video disk. Therefore, video disk advertising would include only material readers wished to see. Will this constitute a large enough dollar volume to support an electronic newspaper industry? If an electronic newspaper does not generate sufficient advertising revenue, then income would have to come from subscription revenue. And if subscriptions were too high, people would instead continue to buy the printed version.

Magazines and books, like newspapers, could be accessed via a television set and video disk. You would probably request the magazine or book by dialing an appropriate telephone number. The material would then be transmitted to your video disk recorder over telephone lines.

But there are problems associated with microelectronic substitutes for magazines and books.

An electronic magazine, like a newspaper, may fail for want of advertising. Again, an electronic magazine would get no advertising revenue if readers could access the stories, bypassing all advertising. With no advertising revenue, the electronic magazine subscription price could, like the newspaper, be so high that people return to the paper and print version.

A book could be transmitted to a video disk as easily as a newspaper or a magazine. You could view the book at your television, with better illustrations than any printed book could offer. But the microelectronic book is unlikely to displace the print and paper version. Why? Because once a book is recorded on a video disk, it can be copied quickly and inexpensively. Just one video recording of each new best selling electronic book might get sold into each community. Friends and neighbors would make copies rather than buy additional recordings. This problem has already plagued the computer programming and cassette music industries.

Computer programs are stored on disks much as electronic books would be. Companies that sell computer programs on disks have devised elaborate schemes to protect disks from being copied, but the problem of "borrowing" has never been solved. Some computer programming companies claim that for every program they sell, ten get copied, and these are ten copies that should have been sales.

Companies that tried to record music on standard one-track magnetic tape cassettes found out how easily cassettes could be copied too. In consequence, cassette sales were never high because there were always more copies than there were sales.

Currently the video cassette industry fights this disk copying problem by recording video cassettes commercially with a special "anti-copy" signal on them. A domestic video recorder (such as the Sony Betamax), which you might buy and attach to your home TV set, will play a protected commercial video cassette, but it will not copy one.

But only time will tell whether this simple anti-copying technique will work. Certainly it prevents the casual movie cassette renter from building a private library of copied movies; but anyone with a good electronics background would probably have little trouble thwarting the anti-copying techniques of the video cassette industry.

We can legislate tough anti-copying laws, and the video recording industry can devise increasingly clever anti-copying protection; but historically the cheats have usually stayed a step ahead. Will they this time? Only time will tell.

But you can copy a printed book on a copy machine. Printed books did not disappear with the advent of copy machines, so why should copying destroy the market for magnetic surface books? The answer is that you must copy printed books one page at a time; copying a magnetic surface book, in contrast, is a single action as simple as copying one page of a printed book. Furthermore, it is frequently more expensive to copy a printed book than to buy another copy. It is invariably cheaper to copy a magnetic cassette or video disk than it is to buy another recording.

Unless some method can be found to stop video cassettes and disks from being copied, electronic book pioneers may well go broke, and publishers will avoid this substitute for the paper and print version. The laser disk may be one solution, since they cannot be copied as easily. You can read a laser disk using low-cost electronics, but it takes much more expensive electronics to write a laser disk.

Let us turn our attention next to new industries that may be created by microelectronics. As an example we choose the "bionic person", because of its obvious emotional impact. But other startling new industries may appear at any time.

Televised fiction has portrayed the bionic person as someone whose strength, speed of reflexes, and sundry other capabilities far exceed the normal human being. Perhaps one day such things will be possible, but already some impressive, albeit less spectacular, innovations are occurring.

At the Illinois Institute of Technology in Chicago, researchers working with Professor

Daniel Graupe have developed an artificial arm that responds to tiny electrical impulses detected in the stump of the lost limb. This in itself is not new, but Professor Graupe's group have significantly refined the process. Using a microprocessor, they analyze signals from three electrodes in order to control movement of elbow, wrist, hand, and fingers, in a lifelike fashion. This gives the amputee's prosthetic limb responses that are equivalent to a natural limb.

The prosthetic limb, at least in theory, can do more than a human limb. The force behind movement in any joint of the prosthetic limb is limited only by the need for a local power supply, the strength of the material from which the limb is constructed, and by the fact that the prosthetic must remain attached to the limb's stub. Someone with an artificial arm will not be able to throw an automobile out of a baseball field, but he or she could contract their artificial fingers into a grip of awesome force. Furthermore, the artificial limb will be able to handle objects that are as hot or corrosive as the construction material will allow, and that will likely exceed the capacity of human flesh.

Electronic hearing is another area where significant developments are occurring. Researchers understand the ear's nervous system well enough to feed it artificially created signals that give profoundly deaf people hearing. This will not be normal hearing, but it is a usable faculty. At the Smith-Kettlewell Institute of Visual Sciences in San Francisco, Dr. Frank Saunders has taken an interesting, but different approach to helping the profoundly deaf. Dr. Saunders has created a belt that is worn in direct contact with the skin. This belt converts sound waves into patterns of vibrations; the wearer learns to interpret these vibrations as though they were sounds. This is not a particularly sophisticated substitute for hearing, and a potential user will need a great deal of training in order to use this belt. But it has helped profoundly deaf children who cannot understand speech through conventional hearing aids and it also serves as a voice feedback for deaf children learning to speak.

Within twenty-five years, electronics could easily develop normal hearing for the most profoundly deaf. In theory, electronics should also be able to give sight to the blind. The retina of a human eye sees in much the same way that electronics of future cameras, as described earlier in this chapter, will capture photographic images. Rods and cones — the seeing elements of the human eye — function in about the same way as semiconductor seeing devices. The retina of the human eye, today, provides more visual information than electronics could hope to match, given the small surface area of an eye's retina. But new electronic developments continuously allow electronics to do more in less space, and it is completely reasonable to expect that within twenty-five years electronics will be able to rival the retina of the human eye. When researchers learn how this electronic information can be transmitted from the electronics of an artificial eyeball to the nerves of the eye socket, they will give sight to the blind.

Electronics opens new prosthetic horizons because the brain controls the body through electric signals. As we learn to detect and interpret these signals, we can build microelectronic devices to do the interpretation and will thus be able to replace any lost limb or faculty.

But the wonders of electronic prosthesis also raise an interesting social question. No hidden problems are associated with the replacement of limbs and faculties lost by injury or accident. But there are definite problems associated with repairing genetic defects. The laws of natural selection have worked against the propagation of defects, since handicapped individuals either did not procreate at all or did so to a very limited extent as compared to the population at large. Yet defects persist. If microelectronics restores the faculties of genetically impaired individuals to the point where they lead normal lives and do their share of procreation, are we perhaps triggering a genetic time bomb? Without legislating who can have children and who cannot, we may be opening up the possibility for future generations to contain few, if any, unimpaired people.

7

Powerful Tools or Powerful Weapons

Anyone who studies the impact of microelectronics and computers on our future society must concern him/herself with places where computers should not be used.

Computers and microelectronics will, in the next twenty-five years, do much to improve the quality of life. Nevertheless, computers should be excluded by legislation from three important applications: the tabulation of election results, the transfer of large sums of money between banks, and the central operations of stock exchanges.

Consider first the tabulation of election results. How simple it is to punch holes in computer cards rather than mark *x*'s in boxes. The computer cards are collected and fed to a computer. Then, presto! Hardly have the polls closed before election results are available for television networks. What's more, the computer is spitting out enough statistics to choke the most avid sports fan.

What is wrong with letting computers count votes? The answer is that it makes vote rigging easier.

The companies that have pioneered the use of computers in vote counting will roar to the defense of their products. They will explain how they guarantee against tampering. Computers can be run in parallel to compare the results. People can spot-check the computer results by hand counting sample ballots. I will argue with none of these safeguards; I will admit them all. I will even assume that the computer manufacturers can produce ten new foolproof safeguards against saboteurs, meddlers, or intruders into the computer counting system. But these are irrelevant arguments. The problem is that computers have vastly reduced the number of people involved in the vote counting process, and computers have removed the actual vote counting procedure from the human eye to the obscurity of a computer's quiet speed.

In order for computer fraud to be detected, someone must detect the fraud; that is a truism. But what if those charged with detecting the computer fraud are themselves in collusion with the fraud perpetrators? What if the political power structure supervising the election is perpetrating the fraud, bribing the programmers, and supplying the verification? Now the computer is working on behalf of the crooks, with the same efficiency that it would otherwise expend on behalf of the voting public.

Opposition groups or splinter factions on the lunatic fringe are not going to manipulate computer vote counting. These people, if they wished to manipulate the ballot process, would have to do so under the watchful eye of incumbents who, if they are not actually running the election, are likely to be watching it very closely. We must watch for incumbents trying to perpetuate themselves illegally.

If the political power structure supervising an election is itself perpetrating a fraud, it can of course perpetrate the fraud with or without computers. One of Lyndon Johnson's earliest political victories, the one that earned him the nickname "Landslide Lyndon", may have resulted from some nonexistent votes getting added into his column in one precinct. But if this fraud did indeed occur, LBJ must have been close to victory anyway. What if he had been losing statewide by a three or four percent margin? Rigging the vote in one or a few precincts to overcome such a large statewide margin would so distort these local results as to beg inquiry. But a computer could be programmed to randomly switch four or five percent of the vote in every precinct — enough to give the winner an apparent comfortable one percent margin without distorting any local count to the point of attracting attention. Ultimately the loser could have recourse to a hand recount, but when the fraud was exposed, the guilty politicians could with little trouble, claim total innocence while one computer programmer disappeared.

The problem, in a nutshell, is that computers massively increase the level of fraud that can be perpetrated per participant because computers will work just as hard for the bad guys as they will for the good guys.

The chances that computers will be used to

rig election results are not very high. But if it occurs, the consequences could be very far reaching. And what do we gain for taking these chances? Using computers to count votes makes the vote counting process a little cheaper and a little faster. Is it worth it? I think not. To give the public election results on the night of the election, rather than the following morning, hardly constitutes an achievement for which it is worth jeopardizing the security of the ballot. And counting votes by hand is not so very expensive. The ballot is the basis for democracy. Let us involve as many people as possible in the voting process, and do everything in our power to make vote rigging harder not easier. The use of computers in ballot counting must be banned.

Turn next to banks. The reckless abandon with which banks are rushing their operations onto computers is cause for grievous concern. Already, virtually all record keeping and data processing is handled by computers, but at least the overall operations can be checked by auditors who have access to paper money and paper results.

Banks have valid reasons for using computers to handle their day-to-day accounting and reporting. Banks would be unable to operate at their present level of efficiency and speed were it not for the computers that keep track of deposits, withdrawals, balances, and other computations associated with the handling of money. But banks have overstepped the line between operating efficiency and fiscal responsibility in their use of computers to transfer large sums of money between banks. This is referred to as "electronic funds transfer".

In order to appreciate the full magnitude of

the problem we are getting into, let us look for a moment at the overall problem of theft from banks.

Banks attract thieves because banks are where the money is. But there are two aspects of bank thefts:

1) The amount of money a thief can steal.
2) The method a thief uses to steal money.

Outsiders who break into a bank know their act will be detected. Therefore, they steal all they can take. The little guy wanders in, pokes a gun in the teller's face, and takes the cash the teller has on hand. This small time bank robber's crime is obviously going to be detected. In fact, it will probably be photographed. The robber doesn't get very much of the bank's assets since a bank does not keep much of its assets in cash, and very little of that cash finishes up in any single teller's money drawer.

More ambitious bank bandits break into the bank vault, or they rob an armored car. Once again the crime has got to be discovered almost immediately, so the bandits take whatever they can get. And once again they do not get much of the bank's assets since they are limited to stealing the cash on hand.

The insider who embezzles money goes about his or her task in a totally different way because he or she can legitimately hope to escape detection, at least for a while. A clerk may manipulate bank balances, transferring funds into favored accounts, out of which he/she or an accomplice can make withdrawals. Working on the inside, the clerk can probably "cook the books" so that a casual audit does not detect the fraud. But a detailed audit will—so the clerk disappears shortly before the detailed audit takes place.

But when a computer starts performing the bookkeeping for a bank, it is just as available to the embezzler as it is to the bank. On the one hand, a computer can stay way ahead of some lowly clerk manipulating account balances; the computer will blow the whistle on this guy very quickly. But conversely, a programmer can add a few unauthorized programs to the computer. Or knowing how the legitimate computer programs work, the programmer may take advantage of some weakness in program logic to manipulate the data as it is entered into the computer. Now the computer is working on behalf of the embezzler.

In all probability, most embezzlers who have used computers to perpetrate their thefts have permanently escaped detection. This is because computers are so much more capable than humans when it comes to accounting. If a programmer uses his computer to embezzle money, and a human accountant tries to catch the embezzlement by performing an audit, I would put my money on the programmer every time.

Don Parker, the well-known computer crime expert at SRI International (formerly Stanford Research Institute), tells the story of the personnel manager (of a company he steadfastly refuses to name) who became intrigued after noticing a magnificent pair of matching Rolls Royce Camargue automobiles that appeared every day in the company parking lot. Investigations identified two programmers in the company's data processing department as the owners of these magnificent vehicles. Following this discovery, the now nervous personnel manager started a full scale investigation of the data processing department. This immediately tipped off the two programmers who, together with their cars, took off for Brazil. In their absence, the investigation of their activities continued. Yet to this day, the company has never been able to find out what the two programmers did. Had they not been so ostentatious as to buy a pair of matching Rolls Royce Camargues, these two programmers would likely have remained among the silent majority of computer criminals — the ones who never get caught and whose actions are probably never even detected.

The programmer-embezzler has a big advantage over the bank clerk since the computer is working on behalf of the programmer. But in both cases, the amount of money the embezzler can steal must be a small percentage of one single bank's assets, and each actual act of embezzlement must involve a sum of money equal to, or less than, a typical transaction. Otherwise, "cooking the books" to hide the embezzlement would become increasingly difficult.

When you cash a check, the transaction usually involves two banks, the bank that issues the check and the bank that cashes the check. Ultimately banks must reconcile debts among themselves since the bank that cashes a check is owed money by the bank that issued the check. Until recently, such interbank money transfers were handled by using some form of paper: high denomination bills, certificates, or whatever. Now banks have turned to electronic funds transfers to settle accounts among themselves, or simply to send money some place on behalf of a depositor. Instead of handing over a piece of paper, a bank will send a computer message that says in effect, "assume you have received the piece of paper and adjust your computer records accordingly."

Now look at the new horizons that have been opened up for thieves and embezzlers. No longer are they restricted by the relative ease with which a single bank branch can be audited, or by the assets of a single bank branch, or by the relatively puny amount of money involved in a single bank transaction. An embezzler who works within the electronic funds transfer system will not be caught unless two banks are audited, the one from which the bogus transfer originated and the one to which the bogus transfer was sent. Covering the crime is also simpler since two separate sets of books are involved, which makes it much harder to spot an obvious inconsistency. And the amount of money involved in each act of embezzlement can increase since it will be measured against the amount of money involved in individual bank-to-bank money transfers. For example, in 1978 Mr. Stanley Rifkin was convicted of stealing ten million dollars in a single bogus bank-to-bank money transfer. A large bank may transmit and receive more than a billion dollars a day using electronic funds transfer; ten million dollars is small potatoes by comparison. An embezzler working within a single bank branch would have a very hard time diverting ten million dollars in a single act and then hiding the crime. Diverting one hundred thousand dollars in a single act would be ambitious; therefore, stealing ten million dollars would require a hundred acts of embezzlement, and that would give the embezzler one hundred times as many chances of getting caught.

I have frequently denounced electronic funds

transfers while lecturing before computer professionals, and in every case someone has leaped to the defense of electronic funds transfer, pointing out that it is almost impossible to decode or alter one of these computer-to-computer messages. And indeed it is, because these messages are "encrypted." Encrypting works much like a scheme that substitutes letters. Here is a simple example of alphabetic encrypting:

For:

ABCDEFGHIJKLMNOPQRSTUVWXYZ

Substitute:

KRLAJQPBYMGXCWHNDOIVEUZFTS

Now the words:

ONE HUNDRED DOLLARS

Become:

HWJ BEWAOJA AHXXKOI

But computers are far more efficient at encrypting than the simple example above illustrates. Computers do not divide data into words; they string the data endlessly as a sequence of ones and zeros. The computer can take any number of consecutive ones and zeros as an arbitrary unit and selectively switch ones to zeros and zeros to ones in any way. In this fashion, a computer can generate ten billion or a hundred billion or a trillion different encrypting codes with equal ease. That means someone tapping into the message may have just one chance in a trillion of breaking the code and messing around with the message. But that is not where the problem lies. The reason banks and computers make such a big deal of their encrypting schemes is because they have found an obvious solution to a trivial problem. The problem, as usual, lies inside the bank. It is an authorized employee transmitting unauthorized messages that causes embezzlement, and there is nothing encrypting can do about that. Moreover, electronic funds transfer has so greatly increased the amount of money that can be embezzled in a single act that this single act may be quite sufficient to keep the embezzler in luxury for the rest of his/her life. If Mr. Rifkin had gotten away with his ten million dollars, for example, he could have lived out the rest of his days quite comfortably, as a law-abiding citizen.

Ah, but Mr. Rifkin was caught, you say; indeed he was. By his own stupidity. He started talking to his friends, who alerted the FBI. But when the FBI went to the bank from which the money had been stolen, the bank still knew nothing of the theft.

For every Mr. Rifkin who has been caught, one hundred equivalent embezzlers have probably gone undetected. In fact, when you look at the rank stupidity that leads to detection of computer criminals, we must wonder with considerable alarm what the clever criminals are doing.

Problems arising from electronic funds transfer do not end with the odd fraudulent bank-to-bank transaction. Thieves always seem to be more imaginative than banks. One of the more celebrated examples of this axiom was an enterprising thief, known professionally as Dr. No. Dr. No set up his own bank on the small island of Sark in the English Channel. The saga of the Bank of Sark is brilliantly recounted, along with other fascinating frauds, in a book called *The Fountain Pen Conspiracy*, by Jonathan Kwitny, a journalist with the "Wall Street Journal". This book is required reading for anyone who thinks that our banks know what they are doing and are secure from embezzlers and outright fraud. When Dr. No founded the Bank of Sark, he had no money in it. But he printed "certificates of deposit", which are the bank-to-bank equivalent of checks and are now being replaced by electronic bank transfers. If you cash a check for $50 at the supermarket, the supermarket gives you $50 and assumes that your bank account has funds to cover your check. Similarly someone could use a certificate of deposit drawn on the Bank of Sark to obtain money from another bank, perhaps in New York. The bank in New York would assume that the Bank of Sark had funds to cover the certificate of deposit. Now, if your private bank account is empty, your $50 check will bounce and the supermarket will come after you. But it never occurred to banks that certificates of deposit would be issued by banks that had no money. And the Bank of Sark had none. Two hundred million dollars of worthless certificates of deposit had been honored by banks all over the world before the fraud was exposed. But the people at the Bank of Sark had disappeared.

So in addition to worrying about simple acts of embezzlement, we must concern ourselves with massive and massively imaginative fraud some criminal genius may perpetrate by manipulating electronic funds transfer.

Within electronic funds transfer the possibility exists for imaginative fraud so massive that by the time it is detected, the perpetrators could be too rich to prosecute. In fact, the moneyed aristocracy of the twenty-first century might be the children and grandchildren of criminals who loot our banking system via electronic funds transfer frauds.

And what does electronic funds transfer gain us? Banks are able to settle accounts between each other instantly rather than within a few days. And they save a little money doing it. Are these small savings worth the massive risk? I think not. And have accounting firms demonstrated their ability to detect fraud once it is perpetrated? They have not demonstrated this ability within single bank branches, let alone taking on the far more formidable problem of bank-to-bank transfers.

I urge legislation outlawing electronic funds transfer and specifically requiring that all bank-to-bank money transfers be made using paper certificates until safeguards against crime have been demonstrated to the satisfaction of a panel of critical industry experts.

I urge that banks be required, by law, to make public all discovered cases of embezzlement, in particular all cases that were perpetrated by the use of computers or electronic funds transfer. Banks tend to hide such events whenever possible on the grounds that the adverse publicity will scare away customers. Perhaps if banks knew that the spotlight would shine on such events, they would show more care in their use of computers and electronic funds transfers.

Electronic stock exchanges are dangerous for exactly the same reason as electronic funds transfers. If all buy and sell orders reaching the New York or American Stock Exchanges were fed to a computer, it is true that the computer could match orders and fix prices, eliminating the hoard of floor traders and specialists who handle these tasks today. Even though these floor traders and specialists have been labeled parasites by some, what do they cost? A lot in absolute dollars, and very little as a percentage of a single day's stock transactions. Yet replacing floor traders with computers again exposes us to the possibility of computer fraud on a scale so massive as to make even electronic funds transfer frauds seem petty. In a typical day, between one billion and ten billion dollars worth of stock may be bought and sold in the U.S.A. There are thousands of buyers and thousands of sellers dealing in thousands of different stocks. Within an electronic stock exchange, who will catch the embezzling stockbroker in a legitimate stockbrokerage house? Who will detect a new Bank of Sark type stockbrokerage? Who can really guarantee that the programmers who put together the central computer system do not have a few crooks in their ranks? No one can, since we can only check for crimes that we have the insight to foresee. Once again, I argue that the small savings that accrue from eliminating floor traders and stock specialists are insignificant compared to the massive potential for fraud and compared to the reckless unknown into which we head by computerizing all stock exchanges. We must legislate to ban computerized stock exchanges. The risk is simply too great, and the reward too small.

The dangers of rushing computers into banks, stock exchanges, and vote counting might perhaps be put into focus if we compared them to the March 1979 Three Mile Island nuclear power plant accident. For many years, nuclear power plant builders assured the public at large that nuclear power plants were completely safe, that every conceivable mishap had been accounted for by building safeguard upon redundant safeguard into every power plant. Perhaps the nuclear power plant designers really believed these assurances; an accident occurred nevertheless. Why? Because the designers could only protect against what they could foresee. What they did not foresee, they could not protect against. That is where problems inevitably occur.

The same is true for computers used in sensitive areas such as electronic funds transfer, vote counting, and stock exchange transactions. Whatever safeguards banks might build into electronic funds transfer, whatever protection the U.S. Securities and Exchange Commission may include in an all-computerized stock exchange, whatever assurances the designers of computerized vote counting may give us, in each case these are human beings whose most honest efforts are only as good as their imaginations. They can all protect us against what they foresee. We must worry about what they cannot foresee. We must ask whether the risks that accompany what they cannot foresee are greater than the economic savings we can all calculate.

The United States Congress has started working on legislation dealing with computer crime and security. As of May 1979, fifteen bills spread among twelve states had been or were being drafted in the area of computer crime. Don Parker of SRI International is concerned that although some of these bills are carefully constructed and will prove effective, others are inappropriate, ineffective, and will be outmoded in five years because their wording is too technically specific to technology of today. Parker, a senior management systems consultant, points out that under the badly worded original draft of Abraham

More than 130 pieces of privacy and security legislation have been introduced in state legislatures since Jan. 1, the Computer and Business Equipment Manufacturers Association has reported.

The bulk of the legislation falls into three categories: criminal records, credit reporting and computer crime.

"The area of computer crime is receiving greater attention in light of the amount of media coverage being given one or two large, but not totally relevant, cases," the Washington-based association said in its state legislation status report for 1979.

Here is a state-by-state rundown of the 15 pieces of computer crime legislation introduced so far this year:

State	Bill Number	Definition of Bill
California	S. 66	Prohibits direct or indirect use of a computer, computer system or network for criminal purposes.
Hawaii	S. 504	Prohibits use of computers for criminal purposes.
Illinois	H. 1027	Makes it illegal to alter computer programs without consent of owner.
Maryland	H. 497	Prohibits fraud by use of a computer and establishes penalties.
Maryland	S. 908	Prohibits fraud by use of a computer.
Massachusetts	H. 4782	Relates to establishing a computer privacy law.
Michigan	S. 848	Prohibits computer fraud.
Missouri	S. 230	Relates to computers, systems, networks, equipment and supplies with penalty provisions.
New Mexico	S. 8	Makes misuse of computers a crime.
North Carolina	S. 397	Makes computer-related crime a felony.
Tennessee	H. 114	Makes unauthorized use of computer equipment a criminal offense.
Tennessee	H. 506	Permits state employees to have access to personal files.
Tennessee	S. 172	Same as H. 114.
Tennessee	S. 514	Same as H. 506.
Utah	H. 183	Prohibits computer fraud.

Computer privacy and security legislation summary, as reported by *Computer World*, May 21, 1979.

A. Ribicoff's Senate Bill S1766, which came before the U. S. Senate in 1976, he was good for forty years in jail. This bill branded as illegal a great deal of the innocent entertainment programmers derive from the computers to which they have access. In this proposed bill of 1976 the same jail sentence was prescribed for the progammer who creates an unauthorized "Snoopy" calendar (a favorite pastime for programmers with time on their hands) and the programmer who executes unauthorized programs to illegally divert bank funds.

A few states (Florida and Arizona), eager to draft legislation against what they saw as an urgent problem, have picked up this original bill and made it law. Yet they will be facing future legal battles as the discretionary powers of law enforcement in the bill are, according to Parker, expanded beyond reason, perhaps to make up for the fact that the bill itself really does not cover computer related crimes well.

At present, however, there is hope. Don Parker feels confident that the California laws covering computer crime should be used as examples for other states. Ribicoff's bill has been revised and modified, reintroduced to the Senate, and will be further revised by the Senate Subcommittee on Criminal Law and by the Department of Justice. It is expected to pass through the Senate in late 1979 and to be passed by the House of Representatives the following year.

8

Silicon — The Shape of Things to Come

Of all jobs held in 1978, perhaps fifty percent could be eliminated during the next twenty-five years. This figure is based on an analysis of data in the "Employment and Earnings" monthly publication put out by the Bureau of Labor Statistics for September 1978.

A pessimist might conclude that if indeed half of all current jobs will be eliminated during the next twenty-five years, then we will ultimately have half the population driving to and from work in armored cars through the other half, which starves while rioting in the streets. But that is an unlikely scenario. If it did happen, the starving fifty percent might just prevail.

Some of the eliminated jobs will be replaced by new jobs. We have said nothing about the industries that exist today but will expand massively during the next twenty-five years; or about indus-

Occupation	Total		Impact*	
	Sept. 1977	Sept. 1978	Percent of Occupation Eliminated	Percent of Total Eliminated
Total employed (thousands)	91,247	95,041		
Percent	100.0	100.0		*53.7*
White collar workers	49.7	49.8		*16.8*
Professional and technical	15.2	14.9	*50*	*8.9*
Managers and administrators, except farm	10.8	10.7	*40*	*4.3*
Sales workers	6.3	6.1	*0*	*0*
Clerical workers	17.4	18.0	*20*	*3.6*
Blue collar workers	33.6	33.8		*28.4*
Craft and kindred workers	13.2	13.4	*90*	*12.1*
Operatives, except transport	11.4	11.6	*90*	*10.4*
Transport equipment operatives	3.9	3.5	*90*	*3.4*
Nonfarm laborers	5.1	5.0	*50*	*2.5*
Service workers	13.6	13.3		*8.5*
Private household workers	1.3	1.2	*0*	*0*
Other service workers	12.4	12.1	*70*	*8.5*
Farm workers	3.1	3.1	*0*	*0*
Farmers and farm managers	1.6	1.6	*0*	*0*
Farm laborers and supervisors	1.5	1.5	*0*	*0*

*Before allowance for new jobs created to fill leisure time resulting from a shorter work week.

A summary of jobs that will be eliminated over the next 25 years. Totals taken from U.S. Bureau of Labor Statistics "Employment and Savings," Vol. 25, No. 10, October 1978. Author's predictions in italics.

tries that do not exist today but will arise out of nowhere.

The industries that are the most immediate and obvious candidates for massive expansion are the telephone and data communications industries. If every home and office is to have one or more computer terminals, and if each terminal spends one or more hours a day connected to a telephone, then the telephone and data communications industries must undergo a truly phenomenal expansion over the next twenty-five years. This increased capacity will be achieved using satellites and microwave communications, together with a greater density of conventional telephone cables. Many of the major computer manufacturers will probably switch their principal business to data communications; IBM is already moving in that direction. We will find IBM, Xerox, AT&T, and ITT locked in competition for this data communications business, with the U. S. Postal Service (if it survives) competing for its share.

The semiconductor industry itself will expand. And the semiconductor industry will spawn thousands of secondary companies that use microelectronics. Robots may displace blue collar production line workers, but someone must design and build the robots.

During 1977 and 1978, the U.S.A. in general, and the state of California in particular, were complaining about high unemployment. But in Silicon Valley, high technology companies experienced a very different problem. They had so many jobs and so few candidates they resorted to stealing each others' employees via radio advertising, and they set up schools to teach the unemployed basic skills needed for low-level positions.

It is inevitable that hundreds or thousands of new companies will be spawned around the semiconductor industry, manufacturing everything from games to robots. How many new jobs will be created? It is hard to say, but compared to the number of eliminated jobs it will not be very high. The number of telephone and data communications industry jobs today totals perhaps one hundred thousand (depending on how jobs get classified). If the size of the data communications industry increases by a factor of ten, that does not mean the number of employees will also increase by a factor of ten. More likely that number of jobs will only double, with microelectronics based automation taking care of the other jobs that might have been.

If 5000 new microelectronics industry companies appear, and each employs approximately one hundred people, then half a million jobs will result.

Existing companies will expand, and they too will create new jobs.

But what will the combined total of new jobs be? A million? Perhaps two million? Even an absurdly high figure cannot begin to approach the fifty million jobs that could be eliminated from the current work force. And the jobs that remain will demand new skills.

For many years to come, therefore, the employment problems facing governments in countries of the industrialized West will be truly formidable. While jobs disappear on a grand scale, there will be persistent labor shortages in electronic and high technology areas. Jobs that are not eliminated will, nevertheless, place severe new demands on job holders. Universities will have to reorganize their curricula to increase the output of engineers, while simultaneously preparing to cope with lifelong learning.

What of the jobless? One solution would be to reduce the work week. Suppose fifty percent of all current jobs were eliminated and ten percent of these eliminated jobs were replaced with new jobs. There is a forty percent net loss of jobs. By reducing the work week from forty hours to twenty-four hours, we could have everyone back to work — assuming that they could handle the new jobs, and that is an improbable assumption.

If everyone were to work twenty-four hours a week, what would they do with the rest of their time? There would be a massive increase in leisure-related occupations that would cater to this new leisure time. Restaurant and hotel businesses would boom. Travel resorts, ski resorts, golf courses, tennis clubs, riding stables, all would see a huge increase in their level of business. There could be a massive cultural renaissance as more people spend more time at the theatre, the symphony, and more time reading.

Alternatively, they might seek second jobs, sleep more, or get drunk or stoned.

But historically, the arts have flourished in societies where even a small minority had enough leisure time to devote to the arts. Professional sports, likewise, must see a massive upsurge in a society where more people have time to spend as spectators. Therefore, manufacturers of sporting goods and equipment will also see good days. It is conceivable that a thirty-five hour work week would create enough new leisure time jobs so that once again we would have full employment.

The microelectronics industrial revolution will generate massive upheavals. In certain industries and geographic locations jobs will be eliminated on a grand scale, and joblessness will be a critical problem, while in other industries and geographic locations there will be a desperate shortage of workers. The politics associated with finding solutions to these problems is, at best, thankless because the net effect is fewer jobs and more unemployment. But if we can adjust to a society where people work less and play more, without reducing productivity, then the future will be very bright.

No one is paying attention to the way in which microelectronics and computers are being used, or to the impact such uses might have on our society. We had better start paying attention, or we will be very, very sorry.

Appendix A
A Worm's-Eye View of Microelectronics

Do you remember what electronic tubes looked like?

Simple electronic tubes used to populate every radio and can still be found in some television sets. At the risk of oversimplification, we can describe an electronic tube as doing for electricity what a check valve does for water. The valve lets water flow in one direction. An electronic tube, likewise, lets electricity flow in one direction only.

Electronic computers were originally built out of electronic tubes — thousands of them. Computer designers had to figure out a way of counting with electronic tubes before they could build computers using them. So they made an "on" tube represent a '1' digit and an "off" tube represent a '0' digit. With that much logic a computer could count from 0 to 1. But computer designers are clever people. With nothing more than 0

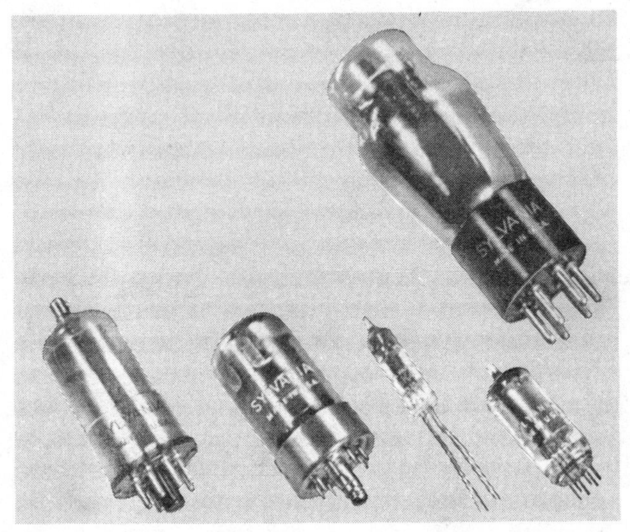

Electronic tubes.

and 1 digits, represented by on and off tubes, they worked out codes to handle the most complex logic or arithmetic.

Even today, the world's largest computer ultimately can do no more than tell the difference between 0 and 1. To solve a problem as simple as adding 2 and 2, a computer must reduce the problem to steps involving only 0 and 1.

But we are not going to explore how computers work. There is no need. Instead, let us look at what an electronic tube can do.

An electronic tube is small; you can hold one in the palm of your hand. An electronic tube is fast; it can switch from on to off, or from off to on, hundreds of times within a single second.

But in the late 1950's electronic tubes were replaced by transistors. Transistors are manufactured using a metal called germanium. Transistors

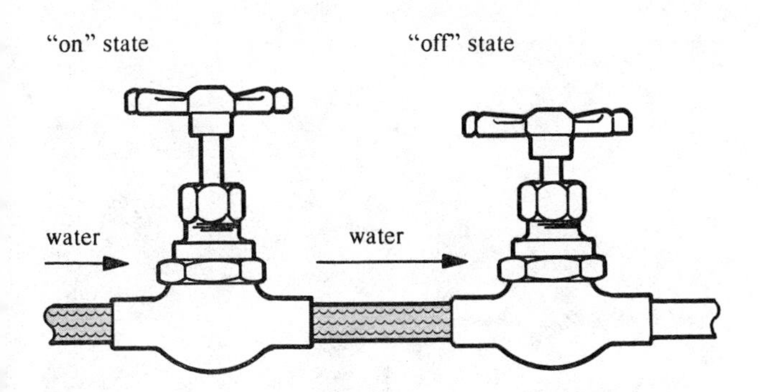

A water check valve lets water flow in only one direction.

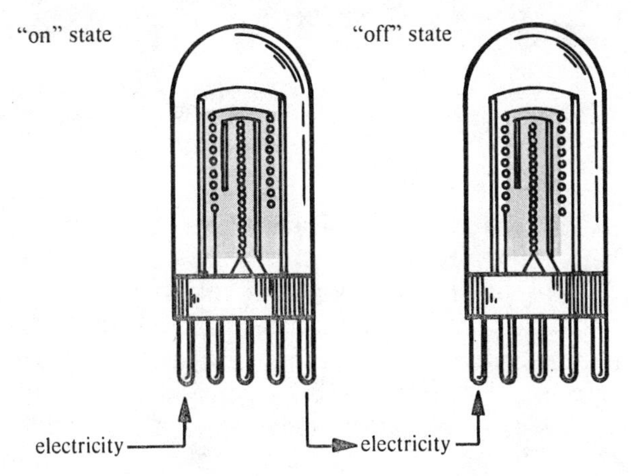

Similarly, an electronic tube lets electricity flow in only one direction.

are a whole lot smaller, a whole lot faster, and a whole lot cheaper than electronic tubes.

Transistors made it easier to build computers, radios, television sets, or anything that had used electronic tubes. Yet no one uses transistors any longer. They use microelectronics.

Photo courtesy of Sperry Univac, a division of the Sperry Corporation.

The world's first electronic digital computer, ENIAC (Electronic Numerical Integrator and Computer). This computer, completed for the U.S. government in 1946, and those designed soon after it, employed electronic tubes by the thousands.

Microelectronic devices are important because they give you so much and cost so little. Perhaps the easiest way of explaining the phenomenal economic impact of microelectronics is to compare microelectronic devices to postage stamps. A sheet of colorful stamps is printed in five steps: a special metallic machine-readable black, and four basic colors, which combine and overlap to generate any pattern or hue. The sheet is then perforated so individual stamps can be separated.

Replace the paper on which the postage stamps are printed with a four-inch diameter silicon disc. Next, replace the postage stamp five-step printing with twelve or more silicon processing steps, some of which are virtually equivalent

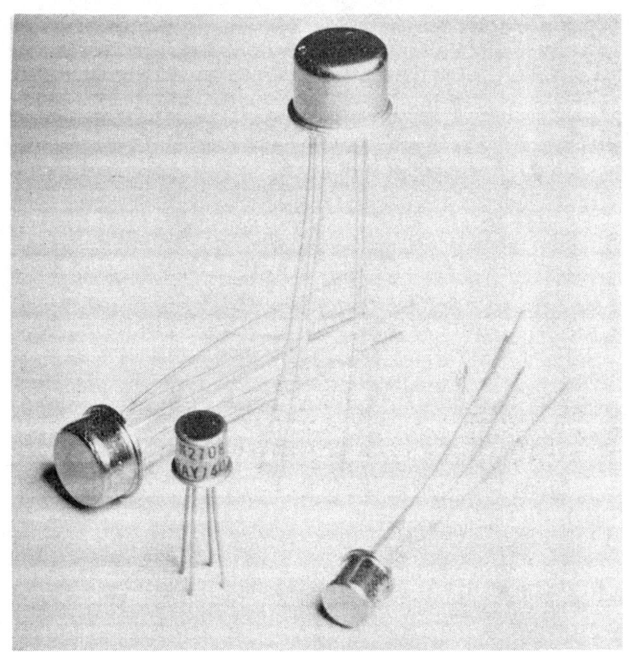

Photo courtesy of Raytheon

The transistor. Size, speed, and cost of the transistor proved a major breakthrough for computer technology.

to printing, and you have a sheet of microelectronic devices that appear to have been printed on the silicon disc. The silicon disk is cut into individual microelectronic devices, just as the sheet of postage stamps is separated into individual stamps. Each microelectronic device is called a "chip".

A silicon disk of microelectronic devices really does look like a shrunken sheet of postage stamps.

Now let us attach some significance to the complexity of a postage stamp's design. Suppose the usefulness of the postage stamp increases as

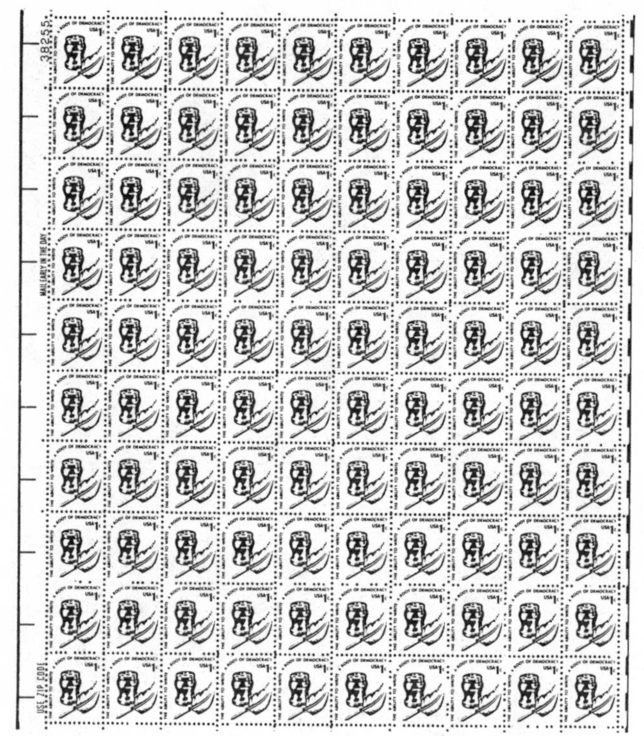

A sheet of postage stamps, photographically reduced in size.

the squares, dots, lines, and curlicues become smaller and more numerous. To make the postage stamps more useful, you might print finer and more complex designs, but ultimately you will reach the limits of fine printing. Lines become so thin they cease to be continuous, bouncing from one microscopic paper hill to the next. Dots, squares, and other solid objects, likewise, will scatter to the point where they cease to be solid objects.

When we switch from postage stamps to

Photo courtesy of Intel Corporation

Silicon discs, printed and sectioned, appear similar to a sheet of postage stamps.

microelectronics, small size and complex design do make microelectronic devices more useful. This is because the electronic tube, which became a transistor, reduces to a pattern created on a piece of silicon. It makes no difference whether the pattern is large enough to fill a silicon chip or microscopically small. In either case the pattern functions similarly. Therefore, as the lines on the silicon become narrower, and the dots smaller, we can fit more lines and more dots into a given area

Graphically, microscopically enlarged paper surface would look like this from above:

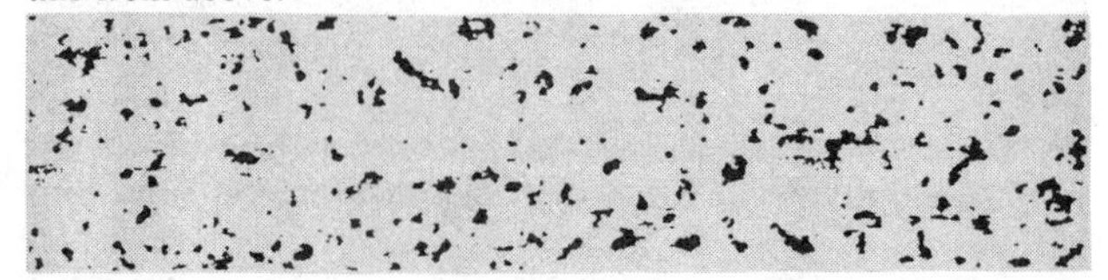

Sideview of same paper surface showing ink deposited onto the "hills" in the paper:

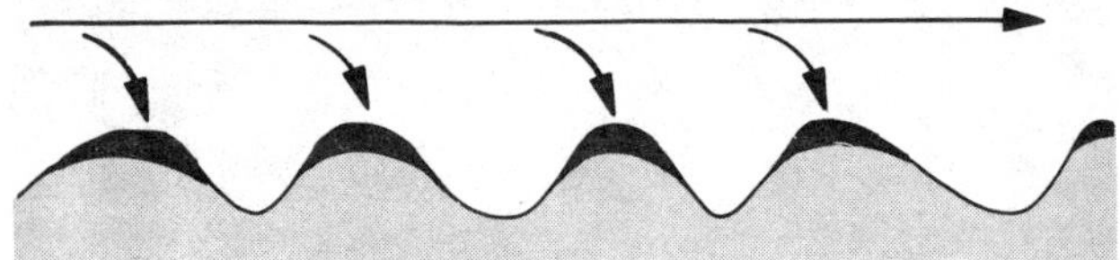

Lines and objects can be printed only so small; smaller, they appear to dissolve. Microscopic "hills" in the paper itself will be colored, while the "valleys" remain blank.

of silicon, and that makes microelectronic devices more useful.

In the beginning, one silicon chip held one large pattern equivalent to a single transistor or electronic tube. That in itself was impressive, since the piece of silicon was much cheaper, smaller, faster, and more reliable than the transistor — which was itself much cheaper, smaller, faster, and more reliable than an electronic tube. But in the past twenty years the printing presses of the semiconductor industry have improved so dramatically that now one silicon chip holds the equivalent of one hundred thousand transistors or electronic tubes. That is to say, while the size of a piece of silicon has not changed, today's pattern printed on that silicon chip is one hundred thousand times as fine as it was twenty years ago.

Silicon chips are fragile. Therefore they are usually housed in plastic or ceramic packages that

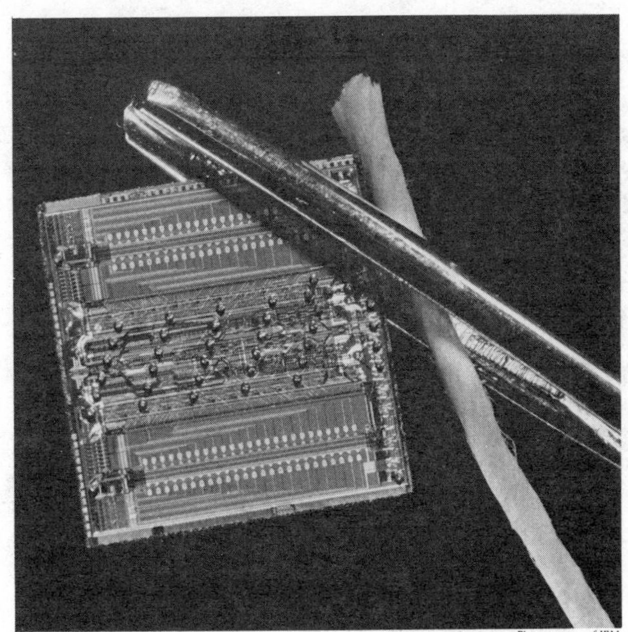

Photo courtesy of IBM.

A silicon chip.

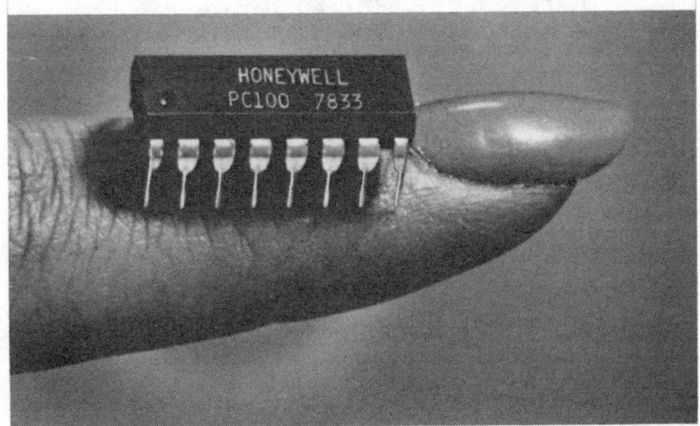

Photo courtesy of Honeywell, Inc.

Fragile silicon chips are housed in plastic and ceramic packages. These packages look rather like insects.

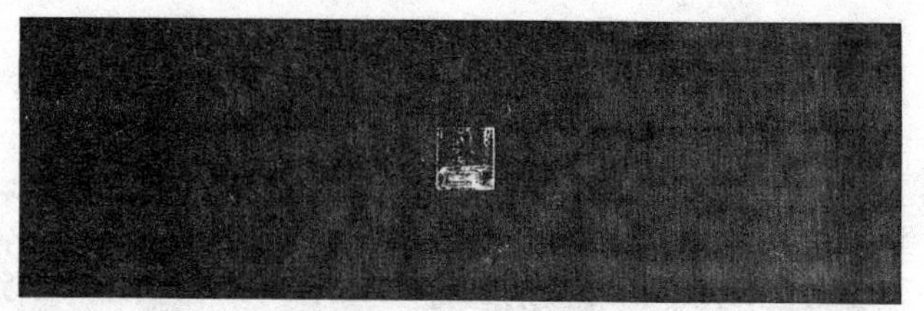

A microcomputer chip, actual size.

look remarkably like insects. Actually, an electronic insect is called (unromantically) a Dual In-Line Package, or DIP for short.

Many different types of electronic logic can be put on a microelectronic chip. One type of logic that is frequently put onto a microelectronic chip is the central processing unit of a computer. This is the "mind" of the computer — the part of the computer that actually makes calculations and executes computer programs. Microelectronic chips with this type of logic are referred to as microprocessors. The most popular microprocessor in the world today is a product manufactured by Intel; the part is named the 8080A.

To return to our comparison, the cost of printing postage stamps does not increase if the stamp design becomes more complex. Nor does the cost of microelectronic chips increase when the designs on the silicon chips become more complex. In fact, the price structure for microelectronic chips has remained remarkably constant through time. When a semiconductor manufacturer first produces his newest whizz-bang chip, he sells the first few for $200 to $300 apiece, because he can get away with it. If the chip is any good, half a dozen companies will be making copies of it within two years, and it will sell for

Photo courtesy of Intel Corporation.

The most popular microprocessor in the world today, Intel's 8080A (greatly enlarged).

two or three dollars apiece. Ultimately the chip will cost one dollar or less. This price structure does not seem to change with time or inflation, which means that chip prices are in effect declining, since we all know that one hundred dollars today buys less than one hundred dollars bought ten years ago.

The electronic tube that cost two dollars in 1950 became a transistor that cost ten cents in 1960. But today we have reduced one hundred thousand transistors to one silicon chip, which if it sells for ten dollars, reduces the price of a single transistor to one hundredth of a cent. That is one healthy rate of deflation. If the dollar had inflated at the same rate at which electronic costs have deflated, a sandwich that cost you a dollar in 1950 would cost you $20,000 today.

As microelectronic chips get cheaper and more dense, they also operate faster. The silicon chip that is equivalent to one hundred thousand electronic tubes operates one million times as fast as the tube. And in the not too distant future, we will compute the time it takes for microelectronic logic operations to occur by measuring distances that electrons must travel within the chip — at the speed of light. Electrons move across the surface of the chip at the speed of light, and the distance the electron moves determines the speed at which the chip operates. Therefore, as the design on the chip becomes finer, and the distances between dots become shorter, the chip operates faster.

And where will it all end? As yet, we do not know. Every two years semiconductor manufacturers can increase by a factor of four the electronics they put on a single silicon chip, and the price remains the same. Therefore, in two years one silicon chip will have on it the equivalent of four hundred thousand electronic tubes, in four years one million six hundred thousand electronic tubes, and in six years, six million four hundred thousand electronic tubes.

Ultimately there is a lower limit to how small microelectronics can get, since the size of individual molecules will become a limiting factor. But long before we reach this limit, problems associated with design and manufacture will occur. Moreover, semiconductor manufacturers are already encountering problems with cosmic radiation. Individual logic units within a microelectronic chip have gotten to be so small that occasional cosmic rays penetrating the earth's atmosphere can change the state of the electronics when they collide with molecular particles.

It is interesting to note that the average microelectronic logic point in 1979 is approximately ten times as large as a single human brain cell. Therefore, when the semiconductor industry increases microelectronic logic density by a factor of ten, it will be dealing with biological scales of size. But microelectronics will still be a long way from competing with biology in terms of what it can accomplish, since a single brain cell is equivalent to perhaps ten thousand microelectronic logic points.

Some people in the semiconductor industry claim that we will indeed encounter an end to the amount of electronics that can be put onto a single chip. But I believe that a number of recent experimental technologies being developed within the semiconductor industry will allow the advance in chip complexity to march on inexorably. If I am wrong, some, but not all, of the predictions in this book will be invalid.

For the more technical readers of this book, the following is a list of technologies I believe will allow the advances in circuit densities to continue:

1) Binary electronic logic will be replaced by quarternary logic or more complex levels of electronics (Dr. Tich Dao of Signetics is probably the "father" of quarternary logic).

2) Three-dimensional semiconductor devices will be manufactured. In other words, a cube of material will be filled with electronic circuits, rather than confining two-dimensional electronics to

the flat surface of a chip. VMOS is a technology in use today that is a step in the three-dimensional direction.

3) New fabrication techniques (such as silicon-on-sapphire) will allow circuits to be packed more closely as well as being finer.

Big numbers are often self-defeating because they become meaningless. After reading too many big numbers, your reaction may well be "so what?" In the case of microelectronics the big numbers quoted mean that future microelectronic devices will be able to do frighteningly complex things, in incredibly short amounts of time, for almost no cost. Therefore, let us give some meaning to the numbers in this book.

One hundred years ago the stagecoach was the principal mode of overland transportation. It moved at approximately twenty-five miles per hour and carried perhaps five passengers. We have come a long way in the last hundred years. Today the Concorde supersonic airliner is the most advanced vehicle of transportation available. It travels at 1300 miles per hour, which is approximately fifty times as fast as the stagecoach. Concorde carries two hundred passengers, which is forty times the capacity of a stagecoach. If Concorde has forty times the capacity of a stagecoach, and fifty times its speed, then in electronic terms, Concorde and the stagecoach are about the same, because in far less time electronics logic capacity has increased by a factor of one hundred thousand, and logic speed of operation has increased by a factor of one million. If Concorde could carry half a million passengers at twenty million miles per hour it would then equal the rate at which microelectronics has advanced in the same time frame. And a ticket for a Concorde flight would have to cost less than a penny if it were to compare with the rate at which microelectronics has gotten cheaper.

Now before you become too impressed with the mind-boggling numbers used to explain microelectronic advances, it is worth noting that the semiconductor industry is, in the end, operated by human beings — with all of their failings. The technology that allowed semiconductor manufacturers to make such impressive gains was frequently more than the human designers could cope with. As each microelectronic product is designed, it is usually given a number not a name. But devices, when first introduced, usually contain so many basic and fundamental design errors that they must be redesigned. The redesigned product then has a letter tacked onto its name. "8080" was the number given to identify the most popular microprocessor on the market today. But people no longer use the 8080. It had too many problems. People now use the "8080A."

Unfortunately, the ability of microelectronic device designers to include glaring errors in their initial designs is matched only by the ability of marketing departments to disguise errors as virtues. Manufacturers produce glossy brochures praising the "features" of a new product to the point where one might define a "feature" as a design error that engineering did not correct so it was dressed up by marketing to seem like a virtue. Yet so incredible are the advances seen in microelectronics that users are prepared to overlook a few "features" in a new product.

Frequently the advances that occur in two or three years can be so stunning that users no longer have sufficient education to understand the new product. An electronic engineer who got his degree ten years ago, if he went back to school today, would understand little or nothing of what is being taught in the classroom. Moreover, the faculties in university electrical engineering departments are years behind industry. No significant developments, either in semiconductor technology or the use of microelectronic products, have come out of universities, and this bleak situation does not look as though it will change in the foreseeable future.

The impressive numbers used to describe microelectronic advances are important because they have turned industrial evolution into industrial revolution.

A technology that is capable of causing an industrial revolution must overwhelm even its own inventors. That is the basic development that causes a situation to run out of control, since no one can understand it well enough to anticipate or plan. And microelectronics is certainly overwhelming its inventors.

Intel Corporation is generally recognized as the technological leader of the semiconductor industry. The Institute of Electrical and Electronic Engineers held a seminar in March of 1976 at which Dr. Robert Noyce, President of Intel, delivered the keynote address. In this keynote address, Dr. Noyce stated that no microelectronic chip would ever be built to equal the complexity of an IBM computer. And yet just two years later his own company introduced a microelectronic chip (it is called the 8086) the complexity of which does nibble at the edges of IBM computers. By 1985 Intel may well be producing microelectronic chips that exactly reproduce the capabilities of IBM computers, and in keeping with the price profile already described they will sell initially for two hundred dollars, and finally for two dollars, as compared to the hundreds of thousands of dollars IBM charges for the same computers today.

If the leaders of the semiconductor industry, such as Dr. Noyce, can predict the pace of the future no better than this, who else stands a chance?

Of course, a single example does not make a case. The case is made by innumerable company presidents who could not see what was happening well enough to get in on the action. Dr. Noyce's poor prediction did nothing to hurt Intel; his own company proved him wrong. But poor predictions of company presidents usually result in new companies springing up to do a job, often by default.

Does it not seem obvious that electronic tube manufacturers would have latched onto transistors, to become the leading transistor manufacturers? Or that the transistor manufacturers would have become the leading semiconductor manufacturers? That was not what happened. Intel, the company I just identified as today's semiconductor technological leader, was founded in 1968 and never built a transistor in its life. The number two semiconductor manufacturer, in terms of technological innovation, is a company called Zilog. Zilog was founded in 1975 by two young engineers who left Intel. They also never built a transistor in their lives.

In fact, few tube manufacturers were smart enough to see transistors coming, and the companies who manufactured transistors, in turn, lost their wits when semiconductors arrived.

Each new technological evolution seems to have initiated corporate revolution.

Appendix B
Glossary

Address Information can be stored electronically. Every electronic storage location is assigned a unique number at the time of manufacture. These numbers become addresses that are used to electronically retrieve the stored information.

Analog A continuous relationship between information and its representation is said to be "analog." For example, a continuously moving watch hand is an analog representation of time. A sound wave is an analog representation of the sound.

Analog recording When sound is recorded as sound waves on magnetic tape, this is referred to as analog recording.

Binary digit (BIT) A binary digit is a number that can have a value of 0 or 1. The number 2 does not exist in the world of binary digits, just as there is no digit beyond 9 in the world of decimal digits. Therefore in the world of binary digits the value 2 becomes 10, just as in the world of decimal digits ten becomes 10. BIT is the commonly used abbreviation for BInary digiT.

Bionics Bionics relates functions and characteristics of living systems to man-made mechanical systems.

Central Processing Unit (CPU) The Central Processing Unit controls every other part of a computer; it is sometimes referred to as the "brains" of a computer. Central Processing Unit is frequently abbreviated to CPU.

Chip A chip is a tiny piece of silicon containing a large amount of electronic logic.

Circuit A circuit is any network of interconnected electronic logic.

Data Base A data base is a large amount of information, stored in computer readable form, with appropriate subject indexes. A computer uses the subject indexes to retrieve specific items of information.

Digital data When information is represented as a sequence of numbers that change in finite steps, the number sequence is referred to as digital data. For example, when time is represented by numbers that identify hours, minutes and seconds, then time is said to be represented digitally, in finite steps of one second.

Digital recording A sound wave may be represented digitally, in which case points on the sound wave, taken at fixed intervals, are each converted into a number. These numbers are recorded in the place of a continuous sound wave.

Dual in-line package (DIP) This is the most widely used packaging for microelectronic circuits. The dual in-line package is a rectangular plastic or ceramic housing with electrical connections divided equally on the two long sides of the rectangle. Dual In-line Package is usually abbreviated to DIP.

Floppy disk A floppy disk is a thin, flexible plastic disk about the size of a 45 rpm record. The disk has magnetic surfaces that are used to store information in computer readable form.

Intelligent terminal An intelligent terminal is a computer terminal consisting of a display screen, a keyboard, and a small computer within the terminal itself.

Laser disk A laser disk is a thin plastic disk whose sides are coated with aluminum. Information is stored on this disk as a pattern of microscopic holes in the aluminum surface. A laser beam is used to create these holes; a lower-powered laser beam is used to detect the presence of holes.

Light emitting diode (LED) A light emitting diode is an electronic component that glows when a small current is passed through it. Light emitting diodes are frequently used to create simple electronic displays, such as digital watch displays which illuminate when you press a button on the watch. Light Emitting Diode is frequently abbreviated to LED.

Liquid crystal display A liquid crystal display uses a thin layer of special liquid between two pieces of glass. The liquid is normally opaque, but in the presence of a tiny current it becomes clear. Patterns of current are used to create patterns of clear liquid that reflect light, thereby generating a display. For example, digital watches that display time continuously have liquid crystals. Liquid Crystal Display is frequently abbreviated to LCD.

Logic Computer programmers use the word "logic" to identify the ideas which they convert into a sequence of instructions that become a computer program. Electronic engineers use the word "logic" to describe the circuits which they create.

Magnetic disk A broader term than "floppy disk," magnetic disks include rigid as well as flexible disks with magnetic coated surfaces that are used to store information in computer readable form.

Mainframe Mainframe is an adjective generally used to describe large computers.

Memory board A memory board is a flat plastic card containing electronic logic that stores addressable information in computer readable form.

Microcomputer A microcomputer is a very small computer that uses a microprocessor for its Central Processing Unit.

Microelectronics Microelectronics is electronic logic represented as microscopic circuits on a chip.

Microprocessor A microprocessor is a single microelectronic chip containing all the electronic logic of a Central Processing Unit (CPU).

Minicomputer Minicomputers are a range of computers supposedly smaller than mainframe computers and larger than microcomputers.

Optical Character Recognition (OCR) This is the process of electronically reading printed characters. Optical Character Recognition is usually abbreviated to OCR.

Personal Computer Industry This industry began building very simple computers for individual hobbyists. The industry now sells microcomputers to all comers, concentrating mostly on small business data processing customers.

Program A program is an explicit sequence of instructions which direct a computer to perform any desired operation.

Programmer A programmer is anyone who writes programs for computers.

Semiconductor A semiconductor is a solid material that conducts electricity under some circumstances but not others. Semiconductors are used in all microelectronic circuits.

Sequential Data Access Accessing data as though it were stored in a long line; using sequential data access you must start at the beginning of the line and work your way down to any point that you wish to reach. Magnetic tape stores data sequentially since you must begin at the beginning of the tape and work your way down it until you reach the information you seek.

Silicon Silicon is a chemical element that is currently the base for almost all semiconductor material.

Silicon Valley Silicon Valley is a slang name for the geographic area between San Francisco and San Jose, California. The primary industry of this area is the manufacture of semiconductors and microelectronic chips out of silicon.

Systems Analyst A Systems Analyst analyzes the job a computer has to do, and defines the job in the form of program logic. A programmer takes this logic and converts it into a computer program.

Appendix C
A List of Companies

The following list provides the names and addresses of companies referred to in the preceding pages.

Alpha Micro Systems
17881 Sky Park North, Suite N
Irvine, CA 92714

Amateur Computer Group of New Jersey
UCTI
1776 Raritan Road
Scotch Plains, NJ 07076

Amdahl Corporation
1250 East Arques Avenue
Sunnyvale, CA 94086

American Electronics Association
2600 El Camino Real
Palo Alto, CA 94306

American Telephone & Telegraph (AT&T)
Bedminster, NJ 07921

Ampex Corporation
401 Broadway
Redwood City, CA 94063

Apple Computer Corporation
10260 Bandley Drive
Cupertino, CA 95014

Atari
2175 Martin Avenue
Santa Clara, CA 95050

Bell Telephone (see American
Telephone & Telegraph)

Bowmar
850 Lawrence Drive
Newbury Park, CA 91320

Brunswick Corporation
One Brunswick Plaza
Skokie, IL 60076

Bulova
75-20 Astoria Boulevard, Jackson Heights
New York, NY 11370

Bureau of Labor Statistics
Western Regional Office
450 Golden Gate Avenue
San Francisco, CA 94102

Burroughs Corporation
1733 N. First Street
San Jose, CA 95112

Casio
P.O. Box 7038
Downey, CA 90241

Centre Electronique Harloge S.A. (CEH)
Rue A. L. Breguet #2
200 Neuchatel
Switzerland

Commodore
3330 Scott Boulevard
Santa Clara, CA 95051

The Computer Store
820 Broadway
Santa Monica, CA 90401

Control Data Corporation
P.O. Box 0
Minneapolis, MN 55440

Cray Research Inc.
1440 Northland Drive
Mendota Heights, MI 55120

Data General
2445 Faber Place
Palo Alto, CA 94303

Datapoint
9725 Datapoint Drive
San Antonio, TX 78284

Diablo Systems/Xerox
24500 Industrial Boulevard
Hayward, CA 94545

Digital Equipment Corporation
146 Main Street
Maynard, MA 01754

Exxon Enterprises, Inc.
1251 Avenue of the Americas
New York, NY 10020

Fairchild Camera & Instrument Corporation
464 Ellis Street
Mountain View, CA 94042

Friden, now TRW Customer Service Division
70 New Dutch Lane
Fairfield, NJ 07006

General Motors Corporation
3044 W. Grand Boulevard
Detroit, MI 48202

Hamilton Watch Company
Lancaster, PA 17604

Heuristics
Box B
N. San Antonio Road
Los Altos, CA 94022

Hewlett-Packard
1501 Page Mill Road
Palo Alto, CA 94304

Honeywell
2025 Gateway Plaza
San Jose, CA 95110

Horn & Hardart
1163 - 6th Avenue
New York, NY 10009

Hughes Research & Development
Microelectronic Products Division
500 Superior Avenue, Department E8
Newport Beach, CA 92663

IBM (International Business Machines)
Old Orchard Road
Armonk, NY 10504

IMSAI (formerly IMS Associates)
14680 Wick Boulevard
San Leandro, CA 94577

Institute of Electrical and Electronic
Engineers (IEEE)
345 East 47th Street
New York, NY 10017

Intel Corporation
3065 Bowers Avenue
Santa Clara, CA 95051

International Computers, Ltd. (ICL)
1 Tutney High Street
London SW15
England

International Telephone & Telegraph (ITT)
320 Park Avenue
New York, NY 10022

Intersil, Inc.
10710 North Tantau Avenue
Cupertino, CA 95014

Lanier Business Products, Inc.
1900 Lafayette
Santa Clara, CA 95050

Lexitron
1400 Coleman Avenue
Santa Clara, CA 95050

Litronix Inc.
19000 Homestead Road
Cupertino, CA 95014

Logical Machine Corporation (LOMAC)
1294 Hammerwood Avenue
Sunnyvale, CA 94086

Magnavox
1700 Magnavox Way
Fort Wayne, IN 46804

Matsushita
Box 288
Osaka, Japan

Mattell
5150 Rosecrans Avenue
Hawthorne, CA 90250

Micro Instrumentation & Telemetry
Systems (MITS)
6328 Linn N.E.
Albuquerque, NM 87108

Microdata Systems
10132 Bilich Plaza
Cupertino, CA 95014

Monroe
P.O. Box 9000R
Morristown, NJ 07960

The Motor Cycle Association of
Great Britain Ltd.
Starley House, Eaton Road
Coventry, England CV12FH

Motorola Incorporated
5005 E. McDowell Road
Phoenix, AZ 85036

National Aeronautics & Space
Administration (NASA)
400 Maryland Avenue SW
Washington, D.C. 20546

National Semiconductor
2900 Semiconductor Drive
Santa Clara, CA 95051

NCR (National Cash Register)
8181 Byers Road
Miamisburg, OH 45409

North Star Computers
(formerly Kentucky Fried Computers)
2547 Ninth Street
Berkeley, CA 94710

Parker Brothers
190 Bridge Street
Salem, MA 01970

Pertec Computer Corporation
P.O. Box 92300
Los Angeles, CA 90009

Popular Electronics Magazine
One Park Avenue
New York, NY 10016

Postal Instant Press, Inc.
P.O. Box 48002
Los Angeles, CA 90048

Qantel Corporation
3474 Investment Boulevard
Hayward, CA 94545

Radio Shack, a division of
Tandy Corporation
1600 One Tandy Center
Fort Worth, TX 76102

RCA
Route 202
Somersville, NJ 08876

Sanders Associates, Inc.
95 Canal Street
Nashua, NH 03060

Seiko Time Corporation
640 - 5th Avenue
New York, NY 10019

Shell Development Company
1 Shell Plaza
P.O. Box 2463
Houston, TX 70001

Signetics
811 East Arques Avenue
Sunnyvale, CA 94086

Singer Corporation
30 Rockefeller Plaza
New York, NY 10020

Smith-Corona Marchant (SCM)
299 Park Avenue
New York, NY 10017

Smith-Kettlewell Institute of
Visual Sciences
2232 Webster
San Francisco, CA 94115

Sony Corporation of America
9 West 57th Street
New York, NY 10019

SRI International
(formerly Stanford Research Institute)
333 Ravenswood Avenue
Menlo Park, CA 94025

Texas Instruments
P.O. Box 225474
Dallas, TX 75265

Timex Corporation
20605 Valley Green Drive
Cupertino, CA 95014

Unimation
Shelter Rock Lane
Danbury, CT 06810

Univac (now Sperry-Univac)
P.O. Box 500
Bluebell, PA 19424

U.S. Securities & Exchange Commission
450 Golden Gate Avenue
San Francisco, CA 94102

Vector Graphic, Inc.
31364 Via Colinas
Westlake, CA 91361

Vydec
9 Vreeland Road
Florham Park, NJ 07932

Wang Laboratories
1 Industrial Avenue
Lowell, MA 01851

Warner Communications
75 Rockefeller Plaza
New York, NY 10020

Westlaw
50 West Kellogg
St. Paul, MN 55165

Xerox Corporation
Stamford, CT 06904

Xerox Palo Alto Research Center (PARC)
3333 Coyote Hill Road
Palo Alto, CA 94304

Zilog, Inc.
10340 Bubb Road
Cupertino, CA 95014

Blocpix, a division of Watson-Manning
972 East Broadway
Stratford, CT 06497

QWIP Systems
Division of Exxon Enterprises, Inc.
5551 Vanguard Road
Orlando, Florida 32809

References

American Electronics Association 1978 Directory. Palo Alto, Calif.: American Electronics Association, 1978.

"Bowmar Instrument Files for Bankruptcy." *Electronics,* February 20, 1979.

Braun, Ernest, and MacDonald, Stuart. *Revolution in Miniature: The History and Impact of Semiconductor Electronics.* Cambridge: Cambridge University Press, 1978.

"Broad Digital-Watch Shakeout Coming." *Electronics,* December 25, 1975.

"C/MOS Watch Kit Aims at Potential $50 Million Market." *Electronics,* January 17, 1972.

Dean, K. J., and White, G. "The Semiconductor Story." *Wireless World,* January 1973.

Freeman, C. *The Economics of Industrial Innovation.* London: Penguin, 1974.

Halacy, D. S., Jr. *Computers, The Machines We Think With.* New York: Dell Publishing Co., 1962.

Hawkes, Nigel. *The Computer Revolution.* London: Thames and Hudson, 1971.

"Intel Drops Digital Watch Line." *Palo Alto Times,* September 8, 1977.

International Computers, Ltd. Annual Company Report, 1967.

Kay, Alan, and Goldberg, Adele. "Personal Dynamic Media." *Computer,* March 1977.

Kwitny, Jonathan. *The Fountain Pen Conspiracy.* New York: Knopf, 1973.

"Liquid Crystal, C/MOS Watch Gets Market Date." *Electronics,* January 3, 1972.

Mansfield, E. *The Economics of Technological Change.* London: Longman, 1969.

Martin, James. *Future Developments in Telecommunications.* 2nd ed. Englewood Cliffs, N.J.: Prentice-Hall, 1977.

Martin, James. *The Wired Society.* Englewood Cliffs, N.J.: Prentice-Hall, 1978.

Motor Cycle Association of Great Britain. "The Motor Cycle Industry in a Nutshell." Starley House, Eaton Road, Coventry, England, 1976 and 1979.

"New IC Market: Electronic Watches." *Electronics,* December 21, 1970.

"$19.95 Watch Coming from T.I." *Electronics,* January 22, 1976.

"1978 Microcomputer Survey." *Interface Age,* vol. 4, issue 1, January 1979.

Pilarski, Laura. "Swiss Make Up for Lost Time." *Electronics,* April 28, 1977.

Sagan, Carl. *The Dragons of Eden: Speculations on the Evolution of Human Intelligence.* New York: Ballantine Books, 1977.

Schneiderman, Ron, and Connolly, Roy. "Semiconductor Firms Fight Watch Duty." *Electronics,* March 18, 1976.

Standard and Poor's Annual Labor Cost Study 1974. "1973 Labor Costs and Per-Employee Sales of Selected Industrial Companies."

Standard and Poor's Industry Surveys. "Labor Costs Special Report." October 12, 1978.

"Surviving a Microcomputer Shift." *Business Week,* April 16, 1979.

Tilton, John E. *International Diffusion of Technology: The Case of Semiconductors.* Washington, D.C.: Brookings Institution, 1971.

"T.I.'s Liquid Crystals Set for Digital Watch." *Electronics,* April 24, 1972.

Walker, Gerald. "LCD Startup Bedevils Watch Firms." *Electronics,* August 18, 1977.

Walker, Gerald. "Timex Goes on the Offensive." *Electronics,* June 23, 1977.

Wang Laboratories Annual Company Report, 1967.